A Text Book Of

COMPUTER INTEGRATED MANUFACTURING

(22658)

Semester - VI

THIRD YEAR DIPLOMA COURSE IN MECHANICAL ENGINEERING GROUP

As Per MSBTE's 'I' Scheme Syllabus

MR. VILAS S. TELI

Diploma Production Tech. (MSBTE)
B.E. Production Engineering (MU)
M.E. Mechanical Engineering (MU)
Visiting Lecture, Agnel Technical College
(Polytechnic), Bandra (W)
Mumbai - 400050.

N4565

Computer Integrated Manufacturing (Sem. - VI)   **ISBN 978-93-889825-71-8**

First Edition : January 2020

© : Authors

The text of this publication, or any part thereof, should not be reproduced or transmitted in any form or stored in any computer storage system or device for distribution including photocopy, recording, taping or information retrieval system or reproduced on any disc, tape, perforated media or other information storage device etc., without the written permission of Authors with whom the rights are reserved. Breach of this condition is liable for legal action.

Every effort has been made to avoid errors or omissions in this publication. In spite of this, errors may have crept in. Any mistake, error or discrepancy so noted and shall be brought to our notice shall be taken care of in the next edition. It is notified that neither the publisher nor the authors or seller shall be responsible for any damage or loss of action to any one, of any kind, in any manner, therefrom.

Published By :

NIRALI PRAKASHAN

Abhyudaya Pragati, 1312, Shivaji Nagar

Off J.M. Road, PUNE – 411005

Tel - (020) 25512336/37/39, Fax - (020) 25511379

Email : niralipune@pragationline.com

➢ **DISTRIBUTION CENTRES**

PUNE

Nirali Prakashan : 119, Budhwar Peth, Jogeshwari Mandir Lane, Pune 411002, Maharashtra

(For orders within Pune) Tel : (020) 2445 2044, Fax : (020) 2445 1538; Mobile : 9657703145

Email : bookorder@pragationline.com, niralilocal@pragationline.com

Nirali Prakashan : S. No. 28/27, Dhyari, Near Pari Company, Pune 411041

(For orders outside Pune) Tel : (020) 24690204 Fax : (020) 24690316; Mobile : 9657703143

Email : dhyari@pragationline.com, bookorder@pragationline.com

MUMBAI

Nirali Prakashan : 385, S.V.P. Road, Rasdhara Co-op. Hsg. Society Ltd.,

Girgaum, Mumbai 400004, Maharashtra; Mobile : 9320129587

Tel : (022) 2385 6339 / 2386 9976, Fax : (022) 2386 9976

Email : niralimumbai@pragationline.com

➢ **DISTRIBUTION BRANCHES**

JALGAON

Nirali Prakashan : 34, V. V. Golani Market, Navi Peth, Jalgaon 425001,

Maharashtra, Tel : (0257) 222 0395, Mob : 94234 91860

Email : niralijalgoan@pragationline.com

KOLHAPUR

Nirali Prakashan : New Mahadvar Road, Kedar Plaza, 1st Floor Opp. IDBI Bank

Kolhapur 416 012, Maharashtra. Mob : 9850046155

Email : niralikolhapur@pragationline.com

NAGPUR

Nirali Prakashan : Above Maratha Mandir, Shop No. 3, First Floor,

Rani Jhanshi Square, Sitabuldi, Nagpur 440012, Maharashtra

Tel : (0712) 254 7129; Email : niralinagpur@pragationline.com

DELHI

Nirali Prakashan : 4593/15, Basement, Agarwal Lane, Ansari Road, Daryaganj

Near Times of India Building, New Delhi 110002 Mob : 08505972553

Email : niralidelhi@pragationline.com

BANGALURU

Nirali Prakashan : Maitri Ground Floor, Jaya Apartments, No. 99, 6th Cross, 6th Main,

Malleswaram, Bangaluru 560 003, Karnataka

Mob : +91 9449043034

Email: niralibangalore@pragationline.com

Note : Every possible effort has been made to avoid errors or omissions in this book. In spite this, errors may have crept in. Any type of error or mistake so noted, and shall be brought to our notice, shall be taken care of in the next edition. It is notified that neither the publisher, nor the author or book seller shall be responsible for any damage or loss of action to any one of any kind, in any manner, therefrom. The reader must cross check all the facts and contents with original Government notification or publications.

niralipune@pragationline.com | www.pragationline.com

Also find us on www.facebook.com/niralibooks

Dedicated to My Loving Parents
Mr. Shivram K. Teli
Mrs. Sushma S. Teli

To Loving Sisters
Bhakti S. Teli
Vindali Raut, Divya Raut

My Teachers
Late S. S. Agarwal
Late P. V. Bhattad

Preface ...

I dedicate this book to the students and staff of the Polytechnic institutions of the Maharashtra state. I am pleased to put forward this book on for **"Computer Integrated Manufacturing"** subject having Course Code No. 22658. This book is written as per new syllabus framed by the MSBTE for the Semester VI of the Diploma in Mechanical Engineering. A successful mechanical engineer must understand each and every aspect of advance manufacturing, which include Computer Aided Design (CAD), Computer Aided Manufacturing (CAM) and Computer Integrated Manufacturing (CIM). The concept and meaning of various terminologies of CAD, CAM and CIM. Must be known to an engineering working in the advance manufacturing industries.

This is self sufficient book. I hope that this book will be welcomed by the students and will certainly help them to improve their performance at the examination and also at campus or service, interviews. It is written as per the course contents of the subject. Every efforts are made to furnish information in an easy to read style, with simple diagram and graded exercise.

My since thanks are to Shri. Dineshbhai furia, Pradeepbhai furia, Shri. Jignesh Furia and whole staff of Nirali Prakashan, especially, Mr. Shashikant Patel, Mr. Ilyas Shaikh, Anagha Medhekar, Chaitali Takle for making these efforts possible.

I also thanks to Miss Tanuja J. Patil for helping me to in writing this book.

Constructive criticism and suggestions from the readers for the improvement of the text book will be greatly appreciated.

Please forward your suggestions at vilasengineering02@gmail.com.

Vilas S. Teli

**Syllabus ...**

1. Introduction to CIM   (Marks 10 - Hours 06)

1.1   **Traditional Product Cycle Diagram:** Role of marketing, R&D, Design, PPC, Quality control and sales departments. Disadvantages and Limitations of traditional product cycle.

1.2   **Current Production Needs:** Production rate, Quality, Accuracy, Repeatability, Flexibility, Survival.

1.3   **CIM:** Concept, Advantages and Benefits of CIM.

1.4   **Elements of CIM:** Computer Aided Design (CAD), Computer Process Planning (CAPP), Computer Aided Manufacturing Control (CAMC), and Computer Aided Business Function (CABF).

1.5   **CAD/CAM/CIM Product Cycle Diagram:** Customer, Marketing, Computer Aided Design (CAD), Computer Aided Process Planning (CAPP), Computer Aided Manufacturing Control (CAMC), Computer Aided Business Function (CABF).

2. Product Cycle Development Through CIM   (Marks 14 - Hours 12)

2.1   **Computer Aided Design (CAD):** Geometric modelling, Finite element analysis and optimization, Evaluation and design review (CAE), Concept of concurrent engineering, and list of software for CAE, Simulation, Automated drafting and generation of report.

2.2   **Computer Aided Process Planning (CAPP):** Concept of CAPP, Structure of processes planning software, Methods of CAPP-variant, Generative. Computerized Material Resource Planning (CMRP), Computerized work scheduling.

2.3   **Computer Aided Manufacturing Control (CAMC):** To generate computer program in machining. Interfacing part program to CNC. Computerized control monitoring and control, Computer Aided Quality Control (CAQC). Programmable Logic Control (PLC), Software list like SCADA etc.

2.4   **Computer Aided Business Functions (CABF):** Enterprise Resource Planning (ERP) - Role of ERP in business, Advantage and Applications of ERP software Material Resource Planning (MRP) – Role of MRP in business, Advantage and benefits. MRP softwares. Customer Relationship Management (CRM) - role of CRM in business, Advantage and Applications. CRM software.

2.5   **Product Lifecycle Management (PLM):** Role of PLM in business, Advantage and Applications. PLM software.

2.6   Supply Chain Management (SCM): Role of SCM in business, Advantage and Applications. SCM software.

3. CIM Hardware, Software, Networking and Data Base Management System (DBMS)   (Marks 12 - Hours 08)

3.1   **CIM Networking:** types of network and its characteristics', applications. Types of network topologies-star, bus and ring topology.

3.2   **Component of Networking:** Application software for CIM, Network software and network hardware.

3.3   **Data Base Management System (DBMS):** Data base types - Hierarchical data base, Network data base, Relational data base, Object oriented data base. Functions of data base management system. Advantages of DBMS.

4. Group Technology and Flexible Manufacturing System   (Marks 12 - Hours 08)

4.1   **Group Technology:** Concept, Basis for developing part families, Part classification and coding with example, concept of cellular manufacturing. Advantages and limitations.

4.2   **Flexible Manufacturing System:** Introduction, Concept, Definition and need, Sub-systems of FMS, Comparing with other manufacturing approaches.

4.3 **Major Elements of FMS:** Workstations, Material handling and storage system, Computer control system and human resources.

4.4 **Classification Based on Flexibility:** Dedicated FMS, Random order.

4.5 **Classification Based on Types of Layouts:** Inline layout type, Rotary layout, Rectangular layout, Loop layout type ladder layout type.

4.6 Applications and benefits of FMS, Advantages and Disadvantages of FMS.

5. Automation (Marks 10 - Hours 06)

5.1 **Automation:** Define, Need of automation, High and low cost automation examples of automations.

5.2 **Elements of automation:** Power source, Control unit and feedback control.

5.3 **Types of automation:** Fixed (Hard) automation, Programmable automation and Flexible automation (Soft), Comparison of types of automation.

5.4 **Strategies in automation:** Simplification specializations of operations, multiple operations, Integration of work stations, Increased flexibility, Automated material handling storage system, Online inspection, Online monitoring, Processes control and optimization, Control of plant operations and computer integrated manufacturing.

6. Robot Technology (Marks 12 - Hours 08)

6.1 **Introduction or robotics:** Definition of robot and robotics, Advantages, Disadvantages.

6.2 **Basic components of robot:** Manipulator, End effectors, Actuators, Sensors, Controller, Processor and software.

6.3 **Robot joints:** Linear, Orthogonal, Rotational, Twisting and revolving.

6.4 **Degree of freedom of robot:** Vertical, radial, Rotational traverse, Wrist pitch, Wrist yaw wrist roll.

6.5 **Actuators:** Mechanical, Hydraulic, Pneumatic and electric.

6.6 **End effectors:** Gripper and types.

6.7 **Robot sensors:** Classification of sensors.

6.8 **Basic configuration of robot:** Cartesian, Cylindrical, Polar (spherical).

6.9 **Applications of robot:** Loading unloading, Material handling, Processing operations, Assembly and inspection.

☝ ☝ ☝

Contents ...

Chapter 1

INTRODUCTION TO CIM

Syllabus

1.1 **Traditional Product Cycle Diagram:** Role of marketing, R&D, Design, PPC, Quality control and sales departments. Disadvantages and Limitations of traditional product cycle.

1.2 **Current Production Needs:** Production rate, Quality, Accuracy, Repeatability, Flexibility, Survival.

1.3 **CIM:** Concept, Advantages and Benefits of CIM.

1.4 **Elements of CIM:** Computer Aided Design (CAD), Computer Process Planning (CPP), Computer Aided Manufacturing Control (CAMC), and Computer Aided Business Function (CABF).

1.5 **CAD/CAM/CIM Product Cycle Diagram:** Customer, Marketing, Computer Aided Design (CAD), Computer Aided Process Planning (CAPP), Computer Aided Manufacturing Control (CAMC), Computer Aided Business Function (CABF).

About this Chapter

At the end of this chapter, students will be able to:

- Explain the traditional product cycle with diagram and show all elements on it.
- Explain advantages and benefits of the given CIM system.
- Explain the given CAD/CAM/CIM product cycle with diagram and show elements on it.
- Compare the given traditional product cycle with its counter CAD/CAM/CIM product cycle.

1.1 TRADITIONAL PRODUCT CYCLE DIAGRAM

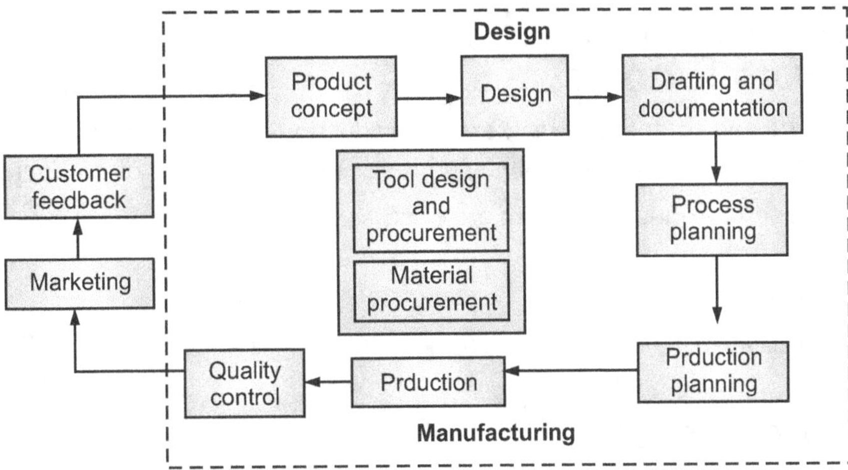

Fig. 1.1 : Traditional Product Cycle

- The various steps in the typical conventional product cycle (Fig. 1.1) are as follows:
    1. **Product Concept:** The product cycle beings with the product concept. This concept is cultivated, refined, analysed and improved.
    2. **Design of Product:** The next step is the design of the product, which include synthesis, analysis and optimization.
    3. **Drafting and Documentation:** The design is followed by its documentation by means of component drawings, assembly drawing, material specifications etc. Here, the design department hands over the bottom to the manufacturing department.
    4. **Process Planning:** The next step is the process planning. Based on the drawings, the process plan is formulated, which specifies the sequence of manufacturing operations. The details specification of tools required, tool layout and material requirement are prepared.
    5. **Production Planning:** This is followed by the production planning of the product.
    6. **Production:** Production planning is followed by the actual manufacturing or production of the product.
    7. **Quality Control:** The product parts are inspected and are made to pass through certain standard quality tests.
    8. **Marketing:** The product is then packaged and shipped to the customers.
    9. **Customer feedback:** The feedbacks from the customers are continuously accepted and incorporated in the design. The product cycle is a continuous closed loop curve.

Advantages and limitations of traditional product cycle:

- Advantages of Traditional product cycle:
    1. Low initial cost of machines; and
    2. Highly trained manpower is not required.
- Disadvantages of Traditional product cycle:
    1. Time required to design and develop the product is high,
    2. Poor design and product quality;
    3. Low rate of production, and hence, low productivity;
    4. High rate of rejection;
    5. Because of low product quality and high product cost, very poor customer satisfaction;
    6. Difficult to optimize the design;
    7. High production cost;
    8. Product modification needs more time.

1.2 CURRENT PRODUCTION CYCLE

- In today's customer driven competitive market, the production system needs to focus on increasing rate of production, improving the product quality and improving the flexibility of production system.
- Various needs of modern production system are discussed below:
    1. High productivity
    2. Low cost of production
    3. High product quality
    4. High Accuracy
    5. Repeatability and Consistency
    6. Flexibility

    1. **High Productivity (Rate of Producton):** The production system must have high rate of production, and hence, high productivity.

2. **Low Cost of Production:** The cost of production should be low. This will help to reduce the cost of the product.

3. **High Product Quality:** The production system should have ability to manufacture the product with high product quality this will lead to better customer satisfaction.

4. **High Accuracy:** The accuracy of the components produced should be high.

5. **Repeatability and Consistency:** The production system should be capable of producing the high quality product consistency.

6. **Flexibility:** The production system should be flexible so as to accommodate the change in design of product with ease in shortest possible time.

1.3 CIM

1.3.1 Introduction to Computer Integrated Manufacturing

- CAD/CAM deals with the integration and automation of the three functions of the factory operations:
  1. Design
  2. Manufacturing planning and control; and
  3. Manufacturing

- However, CAD/CAM does not deal with the business functions of the factory.

- In, earlier days, the business functions were restricted to the management and kept isolated from the factory operations.

- In a present era of globalization and competitive market, the customer is a center of focus. Therefore, for any manufacturing industry, it is necessary to keep on innovating and modifying the product as per the customer's requirement. At the same time, the product should be cost competitive.

- The business decisions are required to be made on regular basis. This is possible only with the integration of business functions and other functions of factory operations. This has led to the emergence of new concept known as Computer Integrated Manufacturing (CIM).

- **Computer Integrated Manufacturing (CIM)** is the complete integration as automation of all functions of factory i.e. design, manufacturing planning and control, manufacturing and business functions.

1.3.2 Concept of CIM

- **Computer Integrated Manufacturing (CIM)** is the complete integration and automation of all functions of factory that are related to manufacturing.

- **Computer Integrated manufacturing (CIM)** system applies computer and communication technology to complete integrate and automate the following four functions of factory operations:
  (i)   Design.
  (ii) Manufacturing planning and control.
  (iii) Manufacturing.
  (iv) Business functions.

- The two terms CAD/CAM and CIM are very closely related. However, the scope and coverage of CIM is broader than that of CAD/CAM.

- CIM includes all functions of factory operations which CAD/CAM covers in addition it also includes business of the factory.

- The scope of CIM is explained in Fig. 1.2.
- The ideal CIM system applies the computer and networking technology to all the operational and information processing functions in manufacturing from other receipt, through design and production, to shipment of product.

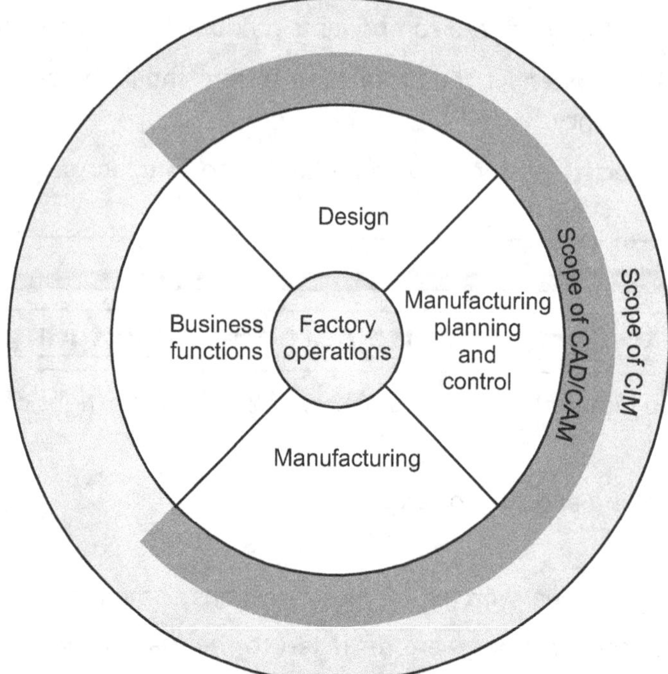

Fig. 1.2 : Scope of CIM

1.3.3 Advantages and benefits of CIM

- CIM plays a vital role in the economy of the manufacturing system is enterprise.
- The benefits of CIM are indicated as follows:

    (i) Products quality improvement.

    (ii) Shorter time in launching new product in the market.

    (iii) Inventory level reduced.

    (iv) Improved scheduling performance.

    (v) Flow time minimized.

    (vi) Competitiveness increases.

    (vii) Shorter Vendor lead time.

    (viii) Improved customer service.

    (ix) Increase in flexibility and responsiveness.

    (x) Long term profitability increases.

    (xi) Total cost minimized.

    (xii) Customers lead time minimized.

    (xiii) Work in process inventory decreases.

    (xiv) Manufacturing productivity increases.

1.4 ELEMENTS OF CIM

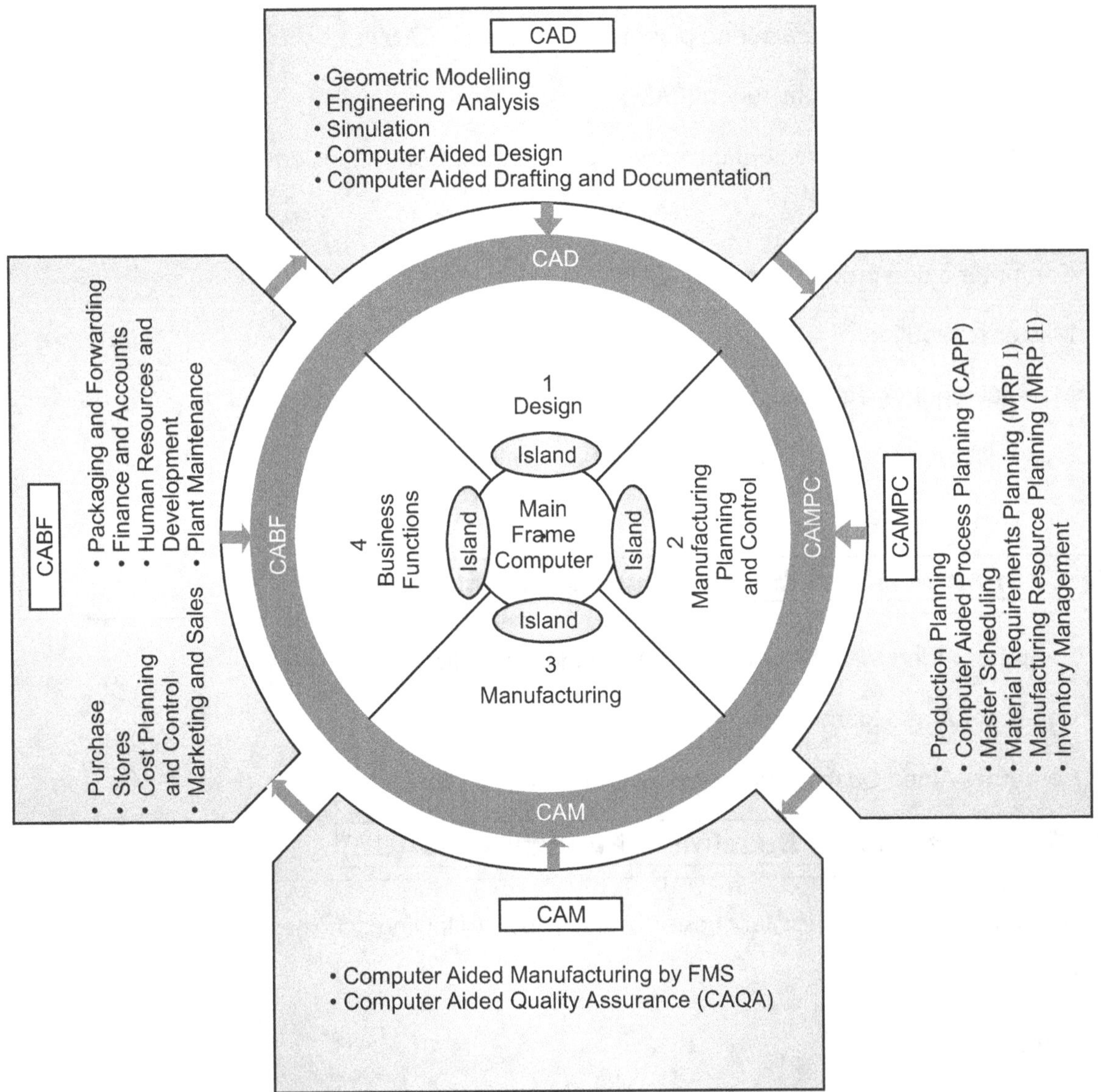

Fig. 1.3 : Computer Integrated Manufacturing (CIM) Wheel

1.4.1 Computer Aided Design (CAD)

- The major activities covered under CAD are:

  ➢ Geometric modelling

  ➢ Engineering analysis

  ➢ Simulation

  ➢ Computer aided design.

  ➢ Computer aided drafting and documentation

1.4.2 Computer Aided Manfacturing Planning and Control (CAMPC)

- The Computer Aided Manufacturing (CAM) is divided into two areas:

    (i) Computer Aided Manufacturing planning and control (CAMPC),

    (ii) Computer Aided Manufacturing (CAM).

- The Computer Aided Manufacturing planning and control (CAMPC) includes the following activity:

    ➢ Production planning.

    ➢ Computer aided process planning (CAPP).

    ➢ Master scheduling.

    ➢ Material requirements planning (MRPI).

    ➢ Manufacturing resource planning (MRPII).

    ➢ Inventory management.

1.4.3 Computer Aided Manufacturing (CAM)

- The Computer Aided Manufacturing (CAM) includes the following activities:

    ➢ Computer Aided Manufacturing by FMS.

    ➢ Computer Aided Quality Assurance (CAQA)

1.4.4 Computer Aided Bussiness Functions (CABF)

- The Computer Aided Bussiness functions (CABF) include following activities:

    ➢ Purchase

    ➢ Stores

    ➢ Cost planning and control

    ➢ Marketing and sales

    ➢ Packaging and forwarding

    ➢ Finance and accounts

    ➢ Human resources and development

    ➢ Plant maintence

- CIM aims to integrate all four islands of automation by interfacing the master computers of all four islands.

- This increases productivity, plant efficient and product quality.

1.5 CAD/CAM/CIM PRODCUT CYCLE DIAGRAM

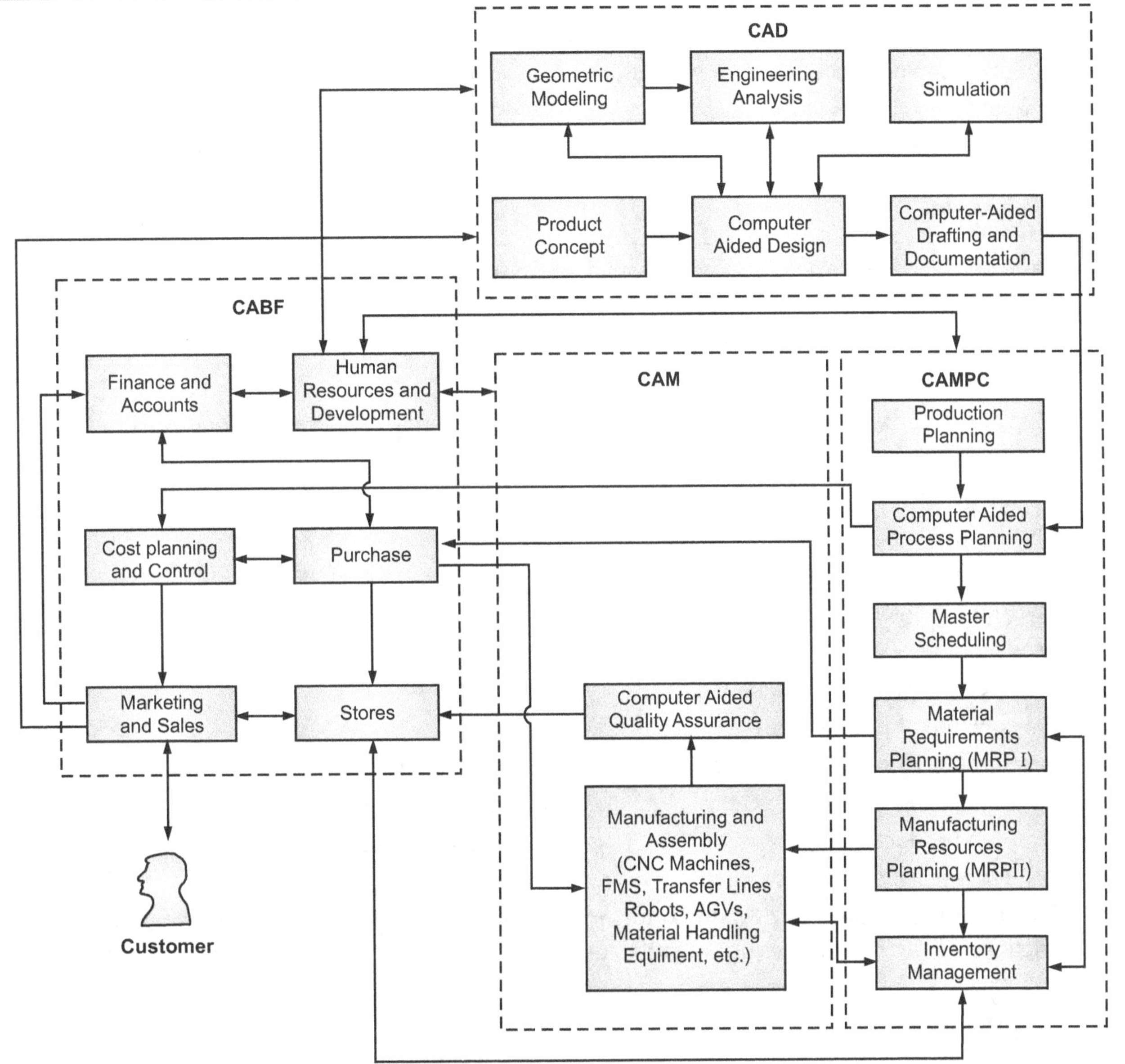

Fig. 1.4 : Step-by-Step Evaluation of Product Cycle with CIM

- The step-by-step evaluation of product cycle with CIM is shown in Fig. 1.4.

- The computer-aided design of the product is carried out using the various CAD tools such as geometric modelling, finite element analysis and simulation.

- From the geometric model, the detail two dimensional drawings of the product and its components are created automatically. This is known as computer aided drafting. At this stage, the complete description and information of the product and its components is available in CAD/CAM database.

- In the next step, the activities are shifted to computer aided manufacturing planning and control (CAMPC). Using the geometric model of the product stored in the database, production planning and computer-aided process planning (CAPP) is carried out.

- CAPP is followed by master scheduling, material requirements plannings and manufacturing resource planning.

- The next step is computer aided manufacturing of the product, using CNC machines, FMS, transfer lines, etc. The computer aided manufacturing is assisted by robots and AGV's for material handling.

- In quality assurance, the computers are used for inspection and performance testing of the product and its components.
- The various departments/activities coming under computer aided bussiness functions are purchase, stores, cost planning and control, marketing and sales, finance and accounts, and human resources and development.
- It is important to note that all activities and functions in CIM are performed by keeping customer at centre point.

Important Points

- Traditional Product Cycle: Various steps involved in traditional product cycle.

  (a) Product Concept

  (b) Design of product

  (c) Drafting and Documentation

  (d) Process planning

  (e) Production planning

  (f) Production

  (g) Quality control

  (h) Marketing

  (i) Customer feedback

- Advantage and Limitation of traditional product cycle:

  Advantage:

  (1) Low initial cost of machine

  (2) High skill labour is not required.

  Disadvantage:

  (1) Poor design and product quality.

  (2) Low rate of production.

- Current production cycle : Various need for current production cycle.

  (a) High productivity

  (b) Low cost of production

  (c) High product quality

  (d) High accuracy

  (e) Repeatability and consistency

  (f) Flexibility

- **CIM (Computer Integrated Manufacturing):** CIM is the complete integration an automation of all functions of factory. i.e. design, manufacturing planning and control manufacturing and business functions.

- **Advantage of CIM**

  (a) Product Quality Improvement

  (b) Inventory Level Reduced

  (c) Improve Scheduling Performance

  (d) Flow time Minimized

  (e) Improve Customer Service

- **Element of CIM**

  (a) CAD

  (b) CAMPC

  (c) CAM

  (d) CABF

Practice Questions

1. Explain traditional product cycle in CIM.
2. State limitations of traditional product cycle.
3. Explain current production need in CIM.
4. Short note on CIM
5. What is the advantages and benefit of CIM?
6. Explain the element of CIM.
7. Explain CIM product cycle diagram.

☝ ☝ ☝

Chapter 2

PRODCUT CYCLE DEVELOPMENT THROUGH CIM

Weightage of Marks = 14, Teaching Hours = 12

Syllabus

2.1 **Computer Aided Design (CAD):** Geometric modelling, Finite element analysis and optimization, Evaluation and design review (CAE), Concept of concurrent engineering, and list of software for CAE, Simulation, Automated drafting and generation of report.

2.2 **Computer Aided Process Planning (CAPP):** Concept of CAPP, Structure of processes planning software, Methods of CAPP-variant, Generative. Computerized Material Resource Planning (CMRP), Computerized work scheduling.

2.3 **Computer Aided Manufacturing Control (CAMC):** To generate computer program in machining. Interfacing part program to CNC. Computerized control monitoring and control, Computer Aided Quality Control (CAQC). Programmable Logic Control (PLC), Software list like SCADA etc.

2.4 **Computer Aided Business Functions (CABF):** Enterprise Resource Planning (ERP) - Role of ERP in business, Advantage and Applications of ERP software Material Resource Planning (MRP) – Role of MRP in business, Advantage and benefits. MRP softwares. Customer Relationship Management (CRM) - role of CRM in business, Advantage and Applications. CRM software.

2.5 **Product Lifecycle Management (PLM):** Role of PLM in business, Advantage and Applications. PLM software.

2.6 Supply Chain Management (SCM): Role of SCM in business, Advantage and Applications. SCM software.

About this Chapter

At the end of this chapter, students will be able to:

- Explain part modelling procedure in CAD for the given component.
- Explain analysis, optimization and evaluation for the given part using any CAE software.
- Explain automated drafting procedure for the given component using any CAD software.
- Differentiate given two methods of CAPP justifying with suitable examples.
- Explain the procedure of computerized part program generation for the given part using any CAM software.
- Explain the procedure of part program interfacing to the given CNC machine.
- Justify the benefits of ERP, MRP, CRM, PLM, SCM using the given corresponding software.

2.1 COMPUTER AIDED DESIGN (CAD)

- Computer Aided Design is an approach to product and process design that utilizes the power of a computer. It involves the computer to create design drawings and product models. CAD covers several automated technologies such as computer graphics to examine the visual characteristics of the product and Computer Aided Engineering (CAE) to evaluate its engineering characteristics.

- Computer aided engineering CAE simplifies the creation of the database to be shared by several applications. These applications include finite element analysis of stresses, strains, deflections, and temperature distribution in structures and load bearing members and generation, storage and retrieval of NC data.

- In CAD, the drawing board and replaced by electronic input and output devices, an electronic plotter. Each section represents a mathematically defined geometric function such as co-ordinate point, line, circle, or cylinder called menu items.

- The CAD system produces quickly accurate models of product and components and the output of the system generates working drawings of very high quality. They can be reproduced number of times and at different levels of reductions and enlargements.

2.1.1 Geometric Modelling

- In geometric modeling is a physical object or any of its parts is represented by geometric model by giving commands that create or modify lines, Surfaces, solids, dimensions and text that together are an accurate and complete two or three dimensional representation of the object.

- The result can be represented in three different ways :

- In line representation, all edges are visible as solid lines. This image can be ambiguous particularly for complex shapes.

- Different colors are generally used for different parts of the object, the making the object easier to visualise. In the Surface model, all the Surfaces are shown but the data describe the interior volume.

2.1.2 Design Analysis and Optimization

- Once the graphic model is created, the design is subjected to engineering analysis. This phase may consist of analysis stresses, stains deflections and other parameters.

- Various sophisticated packages having capabilities to compute this quantity accurately are now available. Because of the relatives ease with which such analysis can be made designers are increasingly willing to throughly analyse a design before it moves on to production.

- Experiments and field measurements may be necessary to determine the effects of loads, temperature and other variables.

2.1.3 Design Review and Evaluation

- An important design stage is review and evaluation to check for any interference between various components in order to avoid difficulties during assembly or use of the part and whether the moving members such as linkage are going to operate as intended.

- The designer can 'zoom' in on part design details and magnify the image on the graphics screen for close scrutiny. Layering is one of the important methods used in CAD design review.

- During design review, the part is precisely dimensioned and tolerenced as required for manufacturing it.

2.1.4 Concept of Concurrnet Engineering

- Market share and profitability are the major determinants of the any organization.

- The factors that influence and improve the competitive edge of a company are unit cost of product quality and lead time.

- Concurrent engineering (CE) has emerged as discipline to help achieve the objectives of reduced cost, better quality and improve delivery performance. CE is perceived as a vehicle for change in the way the products and processes are designed manufactured and distributed.

- Concurrent engineering is a management and engineering philosophy for improving quality and reducing costs and lead time from product conception to product development for new products and product modifications.

- CE means that the design and development of the product, the associated manufacturing equipment and process and the repair tools and processes are handled concurrently.

The concurrent engineering idea contrasts sharply with currently industry sequential practices, where the product is first designed and developed, the manufacturing approach is then established. And finally the approach to repair is determined.

What is Concurrent engineering?

Concurrent engineering is a systematic approach to the Integrated, concurrent design of products and their related processes, including manufacture and support. This approach is intended to cause the developers from the outset, to consider all elements of the product life cycle conception from conception to disposal, including quality, cost, schedule and user requirements.

Concurrent Engineering:

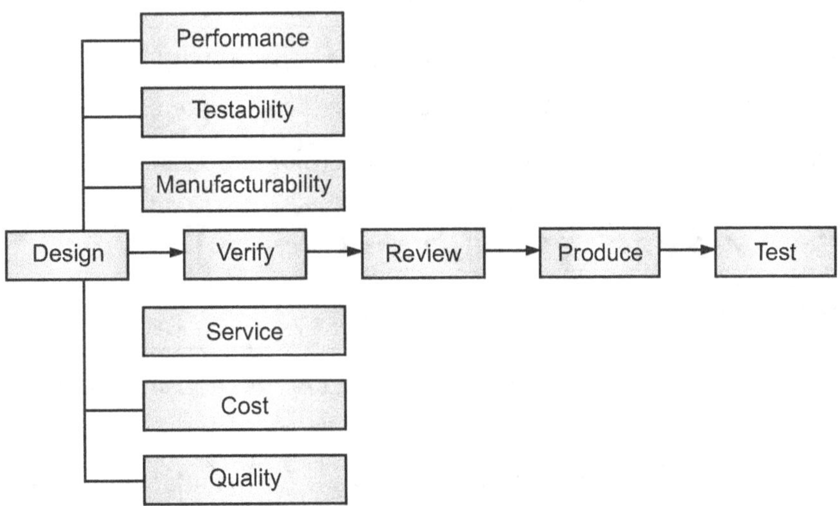

Fig. 2.1 : Concurrent Engineering

2.1.5 List of Software for CAE

1.	MATLAB	2.	Fusion 360
3.	Simulink	4.	Solid Edge
5.	Abaqus	6.	Sim scale
7.	GNU octave	8.	NI multisim
9.	Mathematica	10.	Solid works simulation
11.	Solid work flow simulation	12.	Net logo
13.	Sim events	14.	E-TAP
15.	Simio		

2.1.6 Simulation

- Simulation is the imitation of the operation of real world process, facility or system over a period of time. It involves the generation of an artificial history of a system.

- This artificial history is used to draw the conclusion concerning the operation characteristics of a real system. By developing a simulation model, the behaviour of the system is studied over a period of time

- This model is set of assumptions about operation of the system. Assumptions are usually represented by mathematical, logical and symbolic relationship between the entities, or objects of interest of the system computers are used to evaluate a model mathematically.

- A Simulation model of a system is developed without doubtful assumptions of mathematically solvable models.

- Once model is developed and validated, then it can be used to investigate the desired true characteristics of a real world system.

- For a particular set of input and model characteristic the model is and simulated behaviour is observed. By changing inputs and characteristics, different sets of scenarios are generated and evaluated.

2.1.7 Automated Drafting and Generation of Report

- After analysis and review, the design is reproduced by automated drafting machines for documentation and reference. Detailed and working drawing are developed and printed.

- The automated drafting features include automated dimensioning, scaling of the drawing, development of sectional views and enlarged views of particular part details.

2.2 COMPUTER AIDED PROCESS PLANNING (CAPP)

2.2.1 Concept of CAPP

- Process planning is concerned with concerned with selecting methods of production, toolings, fixtures and machinery sequence of operations and assembly.

- It involves the preparation and documentation, of the plans for manufacturing the products. Computed aided process planning the planning (CAPP) is a means of implementing, the planning function by computer.

- CAPP helps in determining the processing steps required to make a part after CAD has been used to define what is to be made.

- CAPP accomplish these complex steps by viewing the total operation as an integrated system, so that the individual operations and steps involved in making each parts are co-ordinate with others and are performed efficiently and reliably.

2.2.2 Structure of Processes Planning

- Fig. 2.2 represents the structure of a computer aided process system.

- In Fig. 2.2 the modules are not necessarily arranged in the proper sequence but can be based on importance of decision sequence.

- Each module may require execution several times in order to obtain the optimum process plan.

- The input to the system will most probably be a solid model form a CAD data base or a 2-D model.

- The process plan after generation and validation can then be routed directly to the production planning system and production control system.

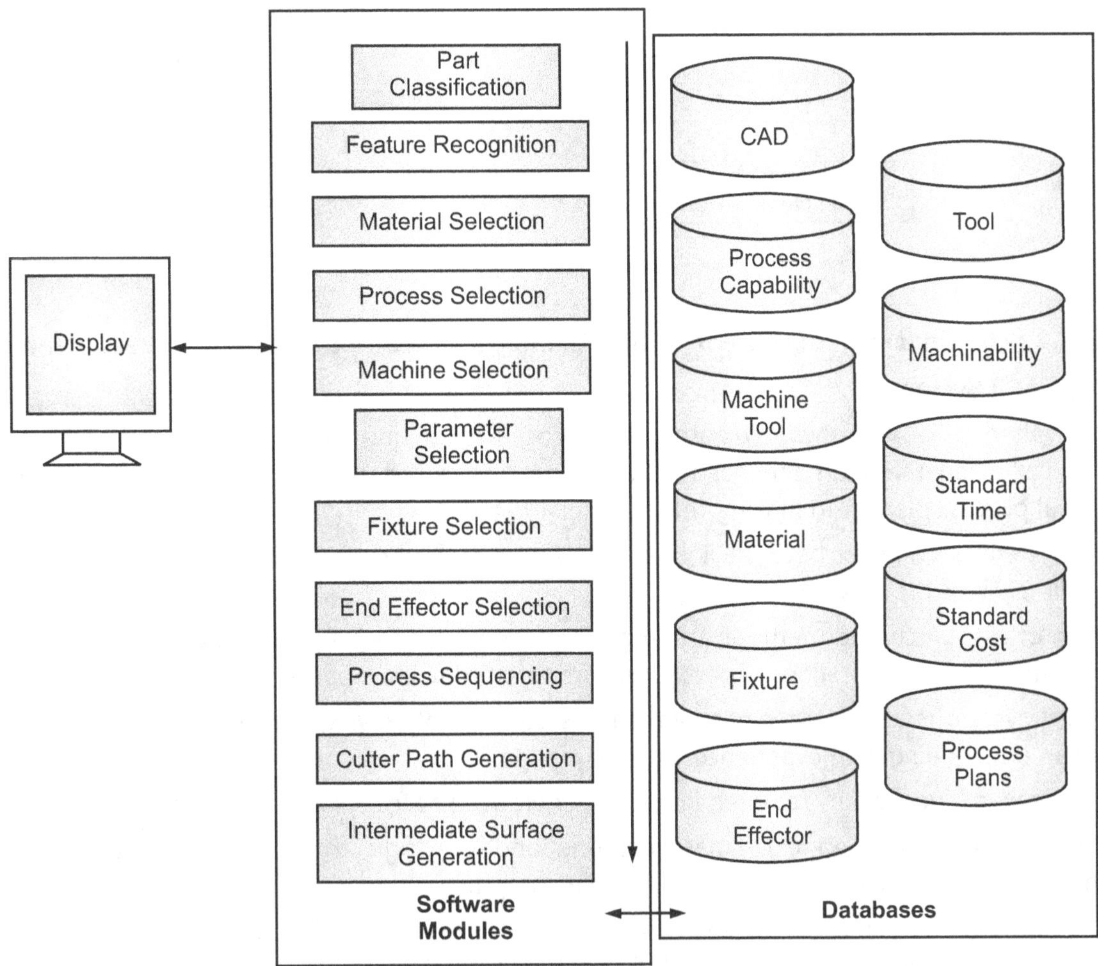

Fig. 2.2 : Structure of a Computer Aided Process Planning System

2.2.3 Method of CAPP-Varient and Generative Process Planning

- There are two types of CAAP systems: Varient and Generative process planning

Varient CAAP System:

- In this system, the computer files contain a standard process plan for the part to be manufactured. The search for a standard plan is then retrieved, displayed for review and printed as a routing sheet.

- The process plan includes operations such as type of tools, and machines to used, the sequence of manufacturing operations to be performed, speeds, feeds, time required for each sequence and so on.

- If the istandard plan for a particular part is not in file of the computer, one that is close it with similar code number and an existing routing sheet is retrieved.

- If a routing sheet does not exist, it is prepared for the new part and stored in the computer memory. The general procedure. For using the retrieval computer aided process planning is shown in Fig. (2.3).

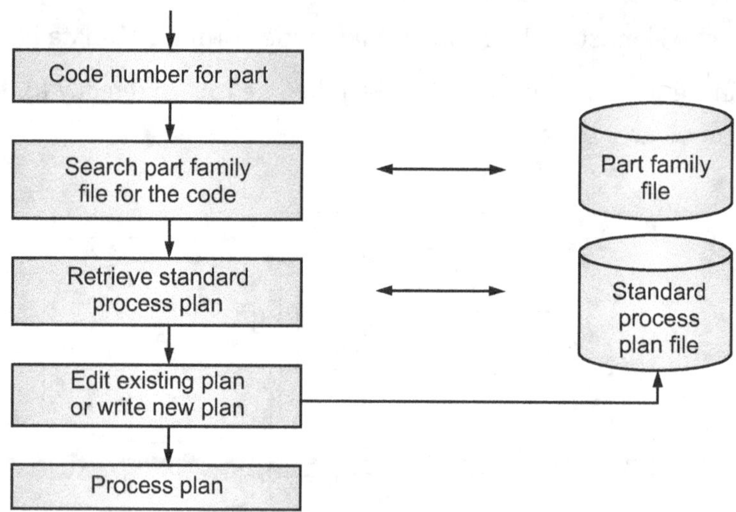

Fig. 2.3 : The General Procedure for Using the Retrieval Computer Aided Process Planning

Generative CAPP Systems:

- It is an alternative approach to automated process planning. In this system, a process plan is automatically generated on the basis of the same logical procedure that would be followed by a traditional process planner in making that particular part.

- Such a system is complex because it must contain comphrehensive and detailed knowledge of part shape, dimensions process capabilities.

- Selection of manufacturing methods and machinery and tools required and sequences of operations to be performed. These capabilities of computers are referred to as expert systems.

- An expert system is a computer program that is capable of solving complex problems that normally require a human being with experience and knowledge.

- There are several ingredients required in a fully generative process planning system.

    (1) The technical knowledge of manufacturing and the logic that is used by successful process planners must be captured and coded into a computer program.

    (2) Computer compatible description of the part to be produced. This description contains all of the pertinent data and information needed to plan the process sequence.

    (3) Capability to apply the process knowledge and planning logic contained in the knowledge base to a given part description.

2.2.3.1 Comparison Between Variant (Retrieved CAPP System and Generative CAPP)

Sr. No.	Varient (Retreiveal) CAPP System	Generative CAPP System
1.	In variant CAPP system, a process plan for a new part is created by user, by modifying existing process plan of a similar parts.	In generative CAPP system the process plan is automatically generated by computer system.
2.	It is manual process.	It is automatic process.
3.	The system required brained and skilled man power.	The system does not require brained and skilled manpower.
4.	It requires large database of existing process plan.	It does not require database of existing process plans.
5.	System does not need high and hardware and software.	System needs high end hardware and software.

2.2.3.2 Benefits of CAPP

(1) Standardization of productivity of process plan.

(2) Standardization of productivity of process planners resulting in reduced lead times, reduced planning costs.

(3) Increases efficient, consistency of product quality and reliability.

(4) Process plan and route can be prepared more quickly.

2.2.4 Computerized Material Resource Planning (CMRP)

- Material requirement planning (MRP) is a computer based inventory. Management system designed to improve productivity for business. Companies use material requirements planning systems to estimate quantities of raw material and schedule their deliveries.

- CMRP works backward from a production plan for finished goods, which is converted into a list of requirements for the subassemblies. Component parts and raw materials that are needed to produce the final products with in the established schedule.

- A critical input for material requirements planning is a bill of material (BOM)-an extensive list of raw materials, components and assemblies required to construct, manufacture or repair a product or service. BOM specifies the relationship between the end product (independent demand) and the components (dependent demand). Independent demand originates outside the plant or production system, and dependent demand refers to components.

- Companies need to manage the types and quantities of materials they purchase strategically; plan which products to manufacture and in what quantities; and ensure that they are able to meet current and future customer demand all at the lowest possible cost. MRP helps companies maintain low inventory levels. Making a bad decision in any area of the production cycle will cause the company to lose money. By mainting appropriate levels of inventory, manufactures can better align their production with rising and falling demand.

2.2.5 Computerized Work Scheduling

- Computerized work scheduling is the process of arranging, controlling and optimizing work and work loads in a production process or manufacturing process through computer control.

- Computerized scheduling is used to allocate plant and machinery resources, plan human resources, plan production processes and purchase materials.

- Purpose of computerized scheduling is to minimize the production time and costs by telling production facility when to make with which staff and on which equipment.

- Computerized work scheduling aims to maximize the efficiency of the operation and reduce costs.

2.3 COMPUTER AIDED MANUFACTURIGN CONTROL (CAMC)

- The manufacturing department takes the database from the process planning department and actually performs the functions of manufacturing of product.

- The computer aided manufacturing and control (CAM) includes the following elements or activities:

    (i) No part programming;

    (ii) Computer aided manufacturing by FMS;

    (iii) Computerized process monitoring and control; and

    (iv) Computer aided quality control (CAQC)

2.3.1 To Generate Computer Programming Machining

- **Computer Aided Manufacturing (CAM):** Compute aided manufacturing can be defined as the use of computer systems to plan, manage and control the manufacturing operations through the direct or indirect computer interface with the manufacturing machine.

Objectives of CAM:

- The basic objectives of computer aided manufacturing are as follows:

    1. Automation and integration
    2. High productivity
    3. High product quality
    4. Reduction in manufacturing lead time
    5. Reduction in manpower
    6. Reduction in material handling
    7. Integration with CAD.

2.3.2 Interfacing Part Program to CNC

- **Part Programing:** The part programming is the set of machining instruction. Written in standard format, for the NC/CNC machine. These instruction can be either punched on the tape using the tape punching machine or directly fed to the computer.

- **Types of part programming:** Based on the method of feeding the part programming to the machine, the part programming can be two types:

    (i) Manual part programming

    (ii) Computers aided part programming

2.3.3 Comptuerized Control Monitoring and Control

- Total computer control fail safe for total computer control system, computers are in complete control of hazardous functions. Usually, the choice to utilize total computer control is made because the task is either very complex or requires a more rapid response to control parameters than can be achieved by non-computerized control.

- The most important characteristics of this category of computerized system is that it is a fail safe compute based control system.

- The use of fail safe control system is limited to those applications in which a computer based control system can be interrupted after a failure occurs without resulting in as impending hazard to space craft or crew.

- The intent of the fail safe concept is to apply validated computer based control system designs that exhibit multiple functionally unique computers (and/or filmware controllers) that reliably can detect the first failure and make the transition of the system to a safe state when failure is detected.

- A control system that is deemed fail safe processes all 10 of the following characteristics:

    (i) A formal development process.

    (ii) Fault containment.

    (iii) Failure and error detection.

    (iv) Controlled system failure detection.

    (v) Failure process.

    (vi) Independence.

    (vii) Prerequisite checks.

(viii) Procedural fault tolerance.

(ix) Hazard detection and safing.

(x) Reconfiguration.

2.3.4 Computer Aided Quality Control (CAQC)

- Because of the number of limitations of manual quality control, use of computer is preferred in quality control.

- Computed aided quality control (CAQC) means the use of computers in automated and on-line quality control functions. The computer aided quality control includes Computer Aided Inspection (CAI) and Computer Aided Testing (CAT).

- The computer aided quality control uses the computers and contact/non-contact type sensors for automatic and preferably on-line 100% inspection.

- In an efforts towards CIM, the CAQC is integrated with CAD/CAM. The computer aided quality control (CAQC) makes the use of CAD/CAM database.

Advantages (Important Features) of Computer aided quality control (CAQC):

- **Some of the advantages of computer aided quality control (CAQC):**
  (i) In CAQC, the inspection during manufacturing is done on-line rather than by taking the part of separate inspection area. This results in saving in time.

  (ii) In CAQC, the data on-line inspection can be used as feedback for control system of machine tool. This will help in maintaining the close tolerances on the dimensions of the component, thereby reducing the rejection.

  (iii) In CAQC, the data of on-line inspection can be stored in computer and used for statiscal analysis of machining processes.

  (iv) In CAQC, the non-contact type sensors are widely used, with non-contact type sensors, the component can be inspected while rotating and without stopping.

  (v) The contact type measuring instruments can cause damage to the surface of finished components. The non-contact type inspection used in CAQC, helps in preventing the damage to the surface of finished components.

  (vi) Unlike manual quality control, there is 100% inspection in CAQC. The 100% inspection helps in a achieving the condition of 'zero defect'.

  (vii) By adopting CAQC, the quality control can be integrated with CAD/CAM which is a step towards CIM.

  (viii) The CAQC takes human factor out of inspection process, thereby improving the quality of product.

Disadvantages of Computer Aided Quality Control (CAQC):

- Though the CAQC is widely accepted for its range of advantages, it has few limitations. The limitations of CAQC are as follows:
  (i) The CAQC requires high initial investment.

  (ii) The CAQC is suitable only for mass production.

2.3.5 Programmable Logic Control (PLC) Software like SCADA etc.

- **Programmable Logic Controller (PLC)** is a microprocessor based device which takes input from the field input devices like : Sensors, switches, buttons etc. and controls the functioning of devices like : relays, actuators, values, motors etc.

- **Programmable Logic Controller (PLC)** is defined as a microprocessor based electronic device that uses a programmable memory for internal storage of instructions and to perform specific functions such as : logic, sequence, timing, counting, and arithematic operations to control the machines.

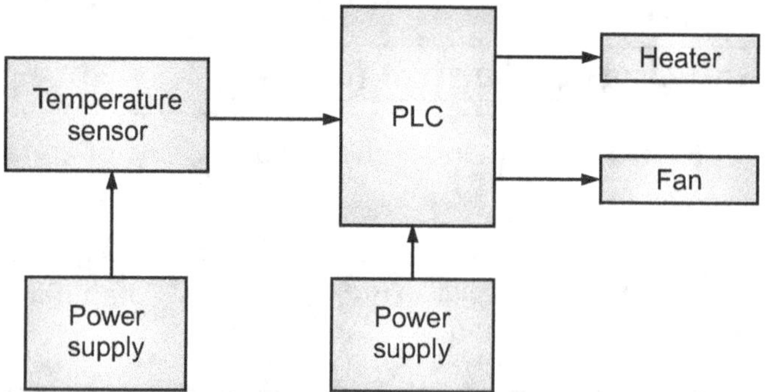

Fig.2.4: Concept of PLC

- Fig. 2.4 shows the concept of PLC. The heater and fan, which are field output devices, are controlled by PLC based on the input from the field input device, temperature sensor. The logic required for control is available in user program, which is stored in PLC memory.

- PLC's were initially intended to replace relay devices.

- **PLC system:** SCADA (Supervisory Control and Data Acquisition) is a computer system used for collecting and analyzing real time data. SCADA systems are plant PLC systems used for monitoring and controlling a plant or equipment is industries such as : chemical process, telecommunications, oil and gas refinery, power generation etc.

2.4 COMPUTER AIDED BUSSINESS FUNCTIONS (CABF)

- The business functions is one of the important island of CIM. At every stage, there is interaction between business function island and remaining three islands.

- The computer aided business functions (CABF) include following elements or activities:
    (i)   Purchase;
    (ii)   Inventory and stock control;
    (iii) Accounting and billing;
    (iv)   Sales order processing;
    (v)   Customer billing;
    (vi)   Packing and forwarding;
    (vii) Marketing;
    (viii) Payroll; and
    (ix)   Plant maintenance

2.4.1 Enterprise Resource Planning (ERP)

- Enterprise Resource Planning (ERP) is a business process management software used for industries or organizations to integrate, automate and manage all the processes and functions needed to run the industry or organization.

2.4.2 Role of ERP in Bussiness

- Various business functions or processes integrated automated and managed by ERP:
    (i)   Planning;
    (ii) Purchase;

(iii)   Inventory control;

(iv)   Accounting;

(v)   Sales;

(vi)   Marketing;

(vii) Customer relations;

(viii) Maintenance;

(ix)   Finance;

(x)   Human resources.

2.4.3 Advantages and Diadvantages of ERP

Advantages of ERP:

1. Real time information is available throughout the company.
2. Better visibility into the performance of operational area of enterprise.
3. Data standardization and accuracy across the enterprise can be achieved.
4. Improves organizational efficiencies.
5. Allows for analysis and reporting for long-term planning.
6. Improves information acess and management.
7. Better monitoring and quicker resolution of queries.
8. Faster response and follow-up on customers.
9. Improved cost control

Disadvantages of ERP:

1. The system can only be customized to a very limited extent.
2. It is rigid and very difficult to adopt to.
3. The system is very expensive to set up.
4. Once the system is set-up switching costs are very high.

2.4.3.1 ERP Softwares

1. Netsuite ERP
2. Sage Intact
3. Dynamics
4. SAP Bussiness
5. ERPAG
6. Tally ERP9
7. Bitrix 24
8. Dolibarr
9. Epicor ERP
10. Win team

2.4.4 Material Resource Planning (MRP)

- Computerized Material Resource Planning (CMPR).

- Computerized Material Resource Planning (CMPR), is a computerized function or computational technique which converts the master schedule for the product into a detailed schedule for the procurement of the raw material and vendor manufactured components.

2.4.5 Role of MRP in Bussiness

1. To ensure right quality of material is available for production at right time to produce right quality of final product.
2. To ensure minimum inventory.
3. To maintain the delivery schedule of the final products.

2.4.6 Advantages and Benefits of MRP

1. Quicker and accurate planning.
2. Issue of intime and correct supply orders.
3. Quicker response to changes in master schedule.
4. Reduction in inventory.
5. Reduction in manpower.
6. Improved customer service.
7. Greater productivity

2.4.7 MRP of Softwares

- MRP of software of the best MRP solutions out there:
  1. Netsuite;
  2. Fishbowl;
  3. IQMS;
  4. JobBoss;
  5. SAP Cloud ERP;
  6. Infor VISUAL and
  7. Odoo

2.4.7.1 Customer Relationship Management (CRM)

- Customer Relationship Management (CRM) is an approach to mange a company's interaction with current and potential customers.

- It uses data analysis about customers history with a company to improve business relationship with customers, specifically focusing on customer retention and ultimately driving sales growth.

- One important aspect of the CRM approach is the systems of CRM that compile data from a range of different communication channels, including a company's website, telephone, email, live chat, marketing materials and more recently social media.

- Through the CRM approach and the systems used to facilitate its bussiness learn more about their target audiences and how to best cater to their needs.

2.4.8 Role of CRM in Bussiness

(1) Improve Customer Satisfaction:

- One of the prime benefits of using a CRM is improving customer satisfaction. By using this strategy, all dealings involving servicing, marketing and selling your products and services to your customers can be conducted in an organized and systematic way.

- You can also provide better services to customers through improve understanding of their issues. For instance, if you need to resolve an issue for a customer, your representatives will be able to retrieve all activity concerning that customer, including past purchases, preferences and anything else that might help in finding a solution quickly.

- In this way, you can receive continuous feedback from your customers regarding your products and services.

(2) Improve Customer Retention (and Revenue):

- By using a CRM strategy for your business, you will be able to improve your customer retention rates. Which often translates into increased revenue for your organization.

- According to Harvard business review, a 5% reduction in your customer defection rate can increase your profits anywhere from 25% - 85%.

- As far as CRM benefits go, this is a golden one. By using the data collected, your team can proactively address at risk accounts as well as reach out to satisfied customers at the right moment, to encourage repeat purchases.

(3) Better Internal Communication:

- Following a CRM strategy, helps in building up better communication within the company.

- Sharing customer data network different departments will enable you to work as a team and help optimize the customer experience-one of several major benefits of CRM.

- Each employee will also be able to answer customer questions on what is going on with their product or services.

- By functioning as well informed team, will help increase the company's efficiency overall and offer a better service-to-services.

(4) Optimize your marketing:

- Another important CRM benefits? It allows you to have a more targeted and cost-efficient marketing program.

- By understanding your customer needs and behaviour, you will be to identify the correct time to promote your product.

- A CRM will also help you segment your customers and give-up insight into which are the more profitable customer groups.

- By using this information, you can set-up you relevant promotions for your groups and execute them at the right time.

- By optimizing your marketing resources in this manner, you give yourself the, best possible chance of increasing your revenue.

(5) Gain Valuable Insights:

- How well is your organization really doing? As a CRM store all the information in one centralized place, this makes it a lot easier to analyze your performance as a whole.

- By pinpointing important information such as revenue generated, leads, as well as results of your marketing companies, you will be able to easily generate reports.

- Better reporting data means you will be able to make effective business decision and improve revenue in the long run.

A CRM also maximize your business performance:

- A CRM system can help maximize your business performance by increasing your upsell and cross-sell opportunities. Up-selling is where you offer customers an upgrade or premium products that are related to their purchase.

- Cross-selling is where you offer complementary products that fall into the same category of their purchase.

- Both these sales strategies can be easily conducted with a CRM, as you will have an understanding about their wants ineeds and patterns of purchase.

- Having this information in a central database means that when an opportunity arises, your sales team can promote as required.

2.4.9 Advantage and Applications of CRM

The major benefits of CRM solution for the business are discussed below:

(1) Bussiness Interactions:

- The CRM encourages good communication among the customers and the business which leads to enhanced customer relationship building with maximum effectiveness.

- The history of customers conversation helps to streamline sales and identify potential customers.

(2) Anytime/Anywhere Information:

- The up-to-date information related to sales history, emails, conversation etc. can be accessed via multiple medium like smartphone, tablet, laptop etc.

- This quickness in accessing the information from anywhere at anytime makes the CRM to be flexible.

(3) Qualifying Leads:

- Every email contact between the customer and the sales professional is streamlined with information regarding who is opening, who is reading etc.

- The integration of e-marketing with CRM system helps to qualify leads.

(4) Opportunity Trends:

- The opportunity management by the CRM system provides insight about future revenue and determine the sales.

- The sales analysis signifies the flow of work future bussiness which results in proactive business.

(5) Define goals:

- Well defined goals allow to enhance the user's productivity. Revelant tools can be used to track the sales process and work flow management.

- CRM is very much useful for identifying the business goals and tracking the information for the same.

(6) Tool Integration:

- CRM supports the integration of outlook mail, Excel spreadsheet, Microsoft word, etc.

- The Excel helps to analyse the data further while the word manages templates and correspondence.

- There are other apt tools integrated for managing email marketing, surveys, social media contacts etc.

(7) Professional Service:

- The CRM from the hands of able professionals, who know the uses of CRM system, helps to provide a more effective solution.

- It is vital to choose the right CRM provider to get the maximum benefit with stability.

- The efficiency of the business and productivity can be improved with good CRM solution. By implementing CRM across all employees, the system is streamlined.

- Customers and sales or lead database can be viewed at anytime which helps to grow the business. One can know the requirements of the customers and serve them better with this knowledge sharing possible via. CRM.

- Customers history can be viewed leads can be qualified automatically, Implementing a CRM is the best solution for managing the business sales and handling customers.

2.4.9.1 CRM Software

1. Zoho CRM
2. Hubspot CRM
3. Salesforce
4. Freshsales
5. Bpm online CRM
6. Pipeline Deals.
7. Act!
8. Sugar CRM
9. Insightly

2.5 PRODUCT LIFE CYCLE MANAGEMENT (PLM)

Product lifecycle management is a system of managing the entire lifecycle of product which includes:

(i)   Inception of product;

(ii) Design of product;

(iii) Manufacturing of product;

(iv) Service of product; and

(v)   Disposal of product

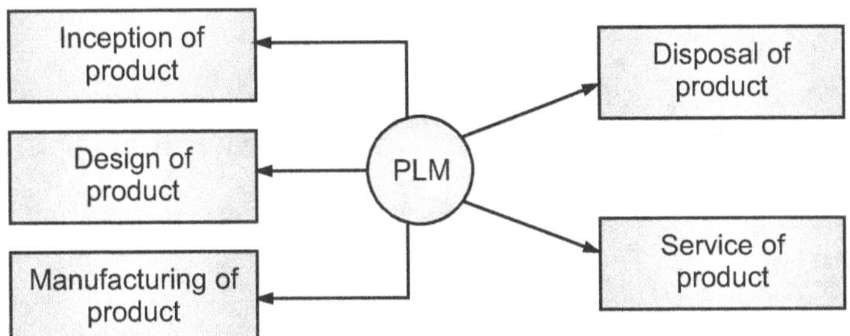

Fig. 2.5 : Product Lifecycle Management

2.5.1 Role of PLM in Bussiness

- In today's competitive environment, where market is flooded with large number of identical products, for a success of a product, it is not only important to minimize the design and manufacturing cost but also necessary to minimize the total product lifecycle cost.

- The product lifecycle cost consist of cost of production, design, manufacturing, service and disposal.

- This has made it necessary to develop a plan to manage all phases of product life cycle, from inception to disposal, at the time of inception/design of product.

- The traditional philosophy of a product design is based on the principle of minimizing the design and manufacturing cost.

- The Product Lifecycle Management (PLM) is a system of managing the entire lifecycle of a product from inception to disposal i.e. from a birth to death of a product.

2.5.2 Advantages and Applications of PLM

(1) Lost Product Lifecycle Cost:

- The 70% of product cost is decided in design stage itself. By proper planning of all phases of product life in design stage itself, the product lifecycle cost can be reduced.

(2) Improve Product Quality:

- By using PLM system, the manufactures can implement all processes necessary to maintain the desired product quality.

(3) Increase Productivity:

- PLM systems avoids/minimizes:
    (i)   time consuming activities;
    (ii)   replication of same data in different parts of systems;
    (iii) Errors and reworks in process.
- This improves the productivity.

(4) Bring Product Innovation:

- The PLM system integrates the efforts of all product development team, irrespective of their geographical locations. This helps to create the best and innovative product design.

(5) Faster Time to Market:

- Because of centralized control over data. The product design can be complied in shortest possible time. In addition, data transfer from design department to manufacturing department is fast and smooth. Therefore,

(6) Minimize Risk of Non-Compliance:

- The PLM system brings uniformity and transparency in organization. This reduces expenses on product recalls and legal issues by improving the compliance.

(7) Increase Earnings of Company:

- Low product lifecycle cost, improved product quality, increased productivity, innovations in product, faster time to market and better compliance results in increase in earning and profit margin of the organization.

2.5.3 PLM Software

1.   Upchair software
2.   Arena
3.   Team center siement
4.   Auto desk vault
5.   Wind chill
6.   Oracle Agile PLM
7.   SAP PLM
8.   Aras PLM
9.   Omnity Empower PLM
10. Propel

2.6 SUPPLY CHAIN MANAGEMENT (SCM)

- Supply chain management is the management of the Flow of goods and services and includes all processes that transform raw materials into final products.

- It involves the active streamlining of a bussiness's supply side activities to maximize customer value and gain a competitive advantage in the marketplace.

- SCM represents an efforts by suppliers to develop and implement supply chains that are an efficient and economical as possible.

- Supply chains cover everything from production to product development to the information systems needed to direct these undertaking.

2.6.1 Role of SCM in Bussiness

The role of global supply chain management primarily comprises five functions mentioned below:

(1) Purchasing:

- This is the first function of supply chain management. It pertains to procuring raw materials and other resources that are required to manufacture the goods.
- It requires co-ordination with suppliers do deliver the materials without any delays.

(2) Operations:

- The operation team engages in demand planning and forecasting. Before giving raw material purchase order, the organization has to anticipate the possible market demand and number of units it needs to produce.
- Accordingly, it further sets the ball rolling for inventory management, production and shipping.
- If the demand is over anticipated, then it could result in excess inventory cost.
- If the demand is under anticipated, the organization wouldn't be able to meet customer demand, thereby leading to revenue loss. So, operations function plays a critical role in supply chain management.

(3) Logistics:

- This function of supply chain management requires immense co-ordination. The manufacturing of products has commenced.
- It needs space for storage until it is shipped for delivery. This calls for making local warehouse arrangements.
- Let's say; the products are to be delivered outside the city, state or country limits. This brings transportation in the loop.
- There will also be a need for outstation warehouses. Logistics ensure that products reach the end-point delivery without any glitches.

(4) Resource Management:

- Any production consumers raw materials, technology, time and labour. However, all the processes need to be efficient and effective.
- This phase is taken care by the resources management function team. If decides the allocation of resources in the right activity at the right time to optimize the production at reduced cost.

(5) Information Workflow:

- Information sharing and distribution is what really keeps all other functions of supply chain management on track.
- If the information workflow and communication are poor, it could break a part the entire chain and lead to mismanagement.

2.6.2 Advantages and Benefits of Supply Chain Management (SCM)

- When a business has an effective supply chain management, it has a competitive advantages in its industry that allows you to decrease the inherent risks when you're buying raw materials and selling products or services. There are many different benefits of supply chain management.
- Here are the seven most important supply chain management benefits:

  (1) Higher Efficiency Rate:

  - When your business is able to incorporate supply chains, integrated logistics, and product innovation strategies, you will be in a great position to not only predict demand as well as act accordingly and this is, without any doubt, one of the main supply chain management.

- Why? When your business implements supply chain management systems, it will be able to adjust more dynamically to the fluctuating economies, emergency markets, and shorter product life-cycles.

(2) Decrease Cost Effects:

- One of the advantages of supply chain management is the cost decrease in different areas. The most important ones are:

  1. Improves your inventory system;
  2. Adjusts the storage space for finished goods which eliminates damage resources;
  3. Improves your system's responsiveness to the actual customer's requirements.
  4. Improves your relationship with both distribution and vendors.

(3) Increase Output:

- One of the main benefits of supply chain management is the communication imporovement.
- This adds upto the co-ordination and collaboration with shipping and transport companies, vendors and suppliers.

(4) Increases your business profit level:

- When you place your business open to the new technologies and an improved collaboration within the different areas, you can be sure that this will ultimately increase your business profit level.

(5) Boost Co-operation level:

- When we are talking about the most successful businesses right now, one of the things they all have in common is the communication.
- In fact, when there is a lack of communication, your vendors and distributions have no idea about what is going on. So, this is definitely one of the main advantages of supply chain management.
- Plus, when you also open your doors and embrance technology, you can also take and advantage of the fact that people do not even need to share the same space in order to be a true communication.
- The communication among the different areas of your business will allow you to have faster access to forecasts, reporting, quotation, statuses, among many other plans in real time.

(6) No more delays in processes:

- One of the main benefits of supply chain management is the fact that through communication, you can actually lower any delays in processes.
- Since everyone is aware of what they are doing as well as what others are doing, this will mitigate any late shipments from vendors, logistical errors in distribution channels, and holds-ups on production lines.

(7) Enhanced Supply Chain Network:

- It is not easy to maintain a sustainable supply chain management system.
- According to some of its advocators, one of the best ways to do it is by using a combination of learn practices (like waste removal, for example) with agile.
- By combining all the information gathered on the different sectors of your business will allow you to have an enhanced supply chain network.

2.6.2.1 Applications of Supply Chain Management (SCM)

Global supply chains pose challenges regarding both quantity and value. Supply and value chain trends include:

- Globalization
- Increased cross-border sourcing.
- Collaboration for parts of value chain with low-cost providers.

- Shared service centers for logistical and administrative functions.
- Increasingly global operations, which requires increasingly global co-ordination and planning to achieve global optimums.
- Complex problems involve also midsized companies to an increasing degree.

2.6.3 SCM Software

1. E2 open;
2. SAP SCM;
3. Logility;
4. Perfect commerce;
5. Oracle SCM;
6. Infor SCM;
7. JDA SCM;
8. Manhattan SCM;
9. Epicor SCM;
10. Dassault systems SCM;
11. Descartes SCM;
12. Highjump SCM;
13. IFs;
14. Watson Supply chain
15. BIujay SCM are some of the examples of best supply chain management software.

Important Points

1. **Computer Aided Design (CAD):**
   - Computer aided design is the use of computers to aid in the creation, modification, analysis or optimization of a design.
   - CAD software is used to increase the productivity of the designer, improve the quality of design, improve communications through documentation, and to create a database for manufacturing.

2. **Geometric Modelling:**
   - In geometric modelling, a physical object or any of its parts is represented by geometric model by giving command that create or modify lines, surface, solids, dimension and text that together are an accurate and complete two or three dimensional representation of the object.

3. **Design Analysis and Optimization:**
   - Once the graphic model is created, the design is subjected to engineering analysis. This phase may consist of analysis stresses, strains, deflections and other parameters.

4. **Design Review and Evaluations:**
   - An important design stage is review and evaluation to check for any interference between various components in order to avoid difficulties during assembly or use of the part and whether the moving member such as linkage are going to operate or intended.

5. **Concurrent Engineering:**
   - It is a management and engineering philosophy for improving quality and reducing costs and lead time from product conception to product development for new products and product modification.

6. Simulation:

- Simulation is the imitation of the operation of a real world process, facility or system over a period of time. It involves the generation of an artificial history at a system.

7. Automated Drafting and Generation of Report:

- After analysis and review, the design is reproduced by automated drafting machine for documentation and reference. Detailed and working drawing are development and printed.

8. Computer Aided Process Planning (CAPP):

- Computer Aided Process Planning (CAPP) is the use of computer technology to aid in the process planning of a part or product, in manufacturing.
- CAPP is the link between CAD and CAM in that, it provides for the planning of the process to be used in producing a designed part.
- Two method of CAPP :
  1. Variant CAPP system.
  2. Generative CAPP system.
- Benefit of CAPP:
  1. Standardisation of productivity of process plan.
  2. Reduce lead time, planning cost.
  3. Increase efficiency, product quality.
  4. Process plan and route can be prepared more quickly.

9. Computerized Material Resource Planning (CMRP):

- Material Required Planning (MRP) is a computer based inventory, management system designed to improve productivity for business. Companies use material requirements planning system to estimate quantities of raw material and schedule their deliveries.

10. Computerized Work Scheduling:

- Computerized work scheduling is the process of arranging, controlling and optimizing work and work loads in a production process or manufacturing process through computer control.

11. Computer Aided Manufacturing Control (CAMC):

- The computer aided manufacturing Control (CAM) includes the following elements or activities:
  1. No part programming.
  2. Computer aided manufacturing by FMS.
  3. Computerized process monitoring and control.
  4. Computer Aided Quality Control (CAQC).

12. Programmable Logic Control (PLC):

- PLC is the microprocessor based electronic device that uses a programmable memory for internal storage of instructions and to performs specific functions such as logic, sequence timing, counting and arithmetic operation to control machines.

13. Enterprise Resource Planning (ERP):

- Enterprise Resource Planning (ERP) is a business process management software used for industry or organizations to integrate; automate and manage all the processes and functions needed to run the industry or organization.

14. Material Resource Planning (MRP):

- Computerized material resource planning a computerized function or computational technique which converts the master schedule for the product into a detailed schedule for procurement of the raw material and vendor manufactured components.

15. Customer Relationship Management (CPM):

- Customer relationship management is an approach to manage a company's interaction with current and potential customers.

16. Product Life-cycle Management (PLM):

- Product life-cycle management is a system of managing the entire life-cycle of product which includes:
    1. Inception of product.
    2. Design of product.
    3. Manufacturing of product.
    4. Service of product, and
    5. Disposal of product.

17. Supply Chain Management (SCM):

- Supply chain management is the management of the flow of goods and service and includes all processes that transform raw materials into final products.

Practice Questions

1. Short note on geometric modelling.
2. Short note on finite element analysis.
3. Short note on concurrent engineering.
4. List the CAE softwares.
5. Short note on simulation.
6. Short note on automatic drafting and generation of report in CIM.
7. Write the Concept of CAPP.
9. Short note on structure of process planning software.
10. Explain CAPP variant method.
11. Explain CAPP generative method.
12. Difference between CAPP variant and generative method.
13. Explain computerized material resource planning.
14. Short note on computerized work scheduling.
15. Explain Computer aided manufacturing control.
16. Explain computer aided quality control.
17. Short note on PLC.
18. Explain the role of ERP in business.

19. Advantage and application of ERP softwares.
20. Short note on material resource planning.
21. Explain role of material resource planning in business.
22. Advantage and benefit of MRP
23. List softwares of MRP.
24. Short note on customer relationship management (CRM).
25. Role of CRM in business.
26. Advantage and application of CRM
27. List the CRM softwares.
28. Short note on product lifecycle management (PLM)
29. Role of PLM in business.
30. Advantage and application of PLM
31. List PLM softwares.
32. Role of supply chain management (SCM) in business.
33. Advantage and application of SCM
34. List SCM softwares.

☝ ☝ ☝

Chapter 3

CIM HARDWARE, SOFTWARE, NETWOKRING AND DATA BASE MANAGEMENT SYSTEM (DBMS)

Weightage of Marks = 12, Teaching Hours = 08

Syllabus

3.1 **CIM Networking:** types of network and its characteristics', applications. Types of network topologies-star, bus and ring topology.

3.2 **Component of Networking:** Application software for CIM, Network software and network hardware.

3.3 **Data Base Management System (DBMS):** Data base types - Hierarchical data base, Network data base, Relational data base, Object oriented data base. Functions of data base management system. Advantages of DBMS.

About this Chapter

At the end of this chapter, students will be able to:

- Explain the given type of network(s) and network topologies with diagram.
- Explain the given application software, network software, and network hardware with its purpose.
- State need of the given DBMS for the specified situation.
- Explain with sketches the given type of database.

3.1 CIM NEWOKING

3.1.1 Types of Network and its Characteristics

3.1.1.1 LAN (Local Area Network)

- Local Area Network (LAN) is privately owned networks covering a small geographic area (less than 1 km), like a home, office, building or group of buildings (example, campus).
- LAN is a group of computers and associated peripheral devices connected by a communications channels, capable of sharing files and other resources among several users.
- LAN transmits data with a speed of several megabits per second (106 bits per second). The transmission medium is normally coaxial cables.
- LAN links computers, i.e., software and hardware, in the same area for the purpose of sharing information.
- Usually LAN links computers within a limited geographical area because they must be connected by a cable, which is quite expensive.

- People working in LAN get more capabilities in data processing and other information exchange compared to stand alone computers.

- People working in LAN get more capabilities in data processing and other information exchanged compared to standalone computers.

- Fig. 3.1 shows simplest form of LAN that connect two computers together.

- A network which consist or less than 500 interconnected devices across several buildings is still recognized as a LAN shown in Fig. 3.2.

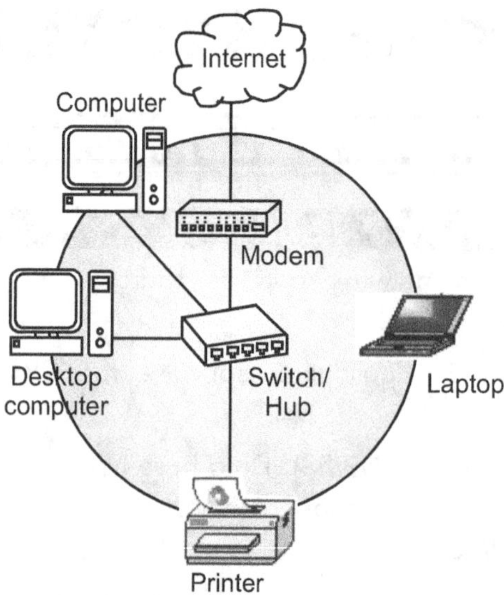

Fig. 3.1 : Local Area Network (LAN)

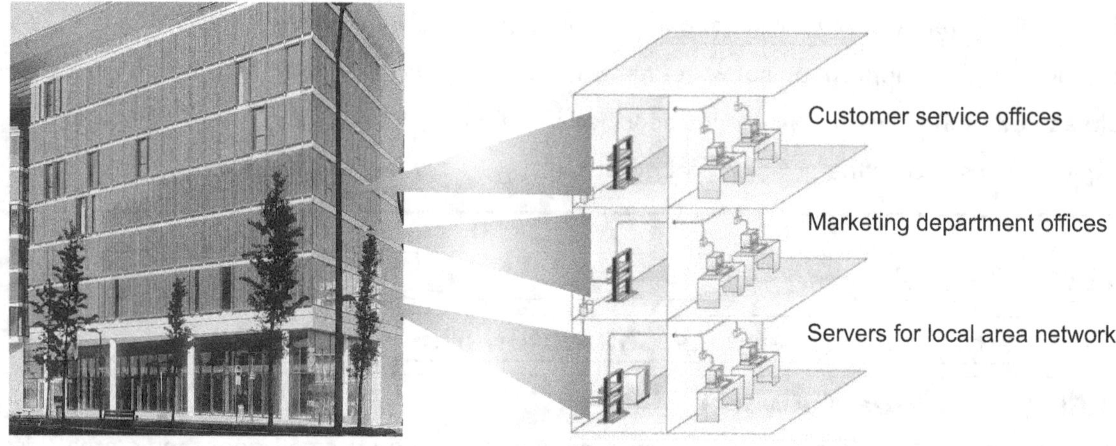

Fig. 3.2

- LAN are widely used for connecting personal computers and workstation in companies offices and factories to share resources.

- Because of this information exchange most of the business and government organizations are using LAN.

- A LAN is a high-speed data network that covers a relatively small geographic area less than 1 km and it typically connects workstations, personal computers, printers servers and other devices.

- LANs offer computer users many advantages, including shared acess to devices and applications, file exchange between connected users and communication between users via electronic mail and other applications.

- LAN have data rate 10 to 100 Mbps.

3.1.1.2 Characteristics of LAN

- Every computer has the potential to communicate with any other computers of the network.
- The reliability of network is high because the failure of the computer in the network does not affect the functioning for other computers.
- High degree of inter connection between computers.
- Easy physical connection of computers in a network.
- Inexpensive medium of data transmission.
- High data transmission rate.
- Less expensive to install.
- Peripheral devices can be shared.

3.1.1.3 Applications of LAN

- Personal computer LANs:
  - ➢ Low cost
  - ➢ Limited data rate
- Backend networks and storage area networks:
  - ➢ Interconnecting large systems (mainframes and large storage devices)
  - ➢ High data rate
  - ➢ High speed interface
  - ➢ Distributed access
  - ➢ Limited distance
  - ➢ Limited number of devices
- High speed office networks
  - ➢ Desktop image processing
  - ➢ High capacity local storage
- Backbone LANs:
  - ➢ Interconnect **Low Speed Local LANs**
  - ➢ Reliability
  - ➢ Capacity
  - ➢ Cost

3.1.2 WAN (Wide Area Network)

- Wide Area Network is a long-distance communications network that covers a wide geographic area, such as a state or country or even the whole world.
- A wide area network is a network that spans a large geographical area, the most common example being the Internet.
- A WAN is contrasted to Smaller Local Area Networks (LANs) and Metropolitan Area Networks (MANs). LANs are home or office networks, while a MAN might encompass a campus or service residents of a city wide wireless or wifi network.

- The internet is a public WAN, but there are many ways to create a business model or private WAN. A private WAN is essentially two or more LAN's connected to each other.

- A WAN provides long-distance transmission of data, voice, image and video information over large geographical areas that may comprise a country, or even whole world.

- WAN containers a collection of machines intended for running user programs we will follow traditional usage and call there machine host.

- A geographically distributed network composed of Local Area Networks (LANs) jointed into a single large network using services provided by common carriers.

- Wide Area Networks (WANs) are commonly implemented in enterprise networking environments in which company offices are in different cities, states or countries or on different continents. WANs span more than one geographical area and are used to connect remote offices to each other.

3.1.2.1 Applications of WAN

- E-mail, conferencing, document exchange, remote database access.

- LAN-to-LAN connections connect two or more geographically separate locations.

- Transaction acquisition the instant relay of transaction information from a point of purchase sale.

3.1.2.2 Characteristics of WAN

- **Areas of Coverage:** WANs located within a country side and worldwide networks, (such as a city, country, or the world) using a communications channel that combines may types of media such as telephone lines, cables and ratio waves. The internet is the world's largest WAN.

- **Distance:** WAN's span (cover) distance greater than 100 miles.

- **Ownership:** WAN's have no ownership.

3.1.2.3 Types of Network Topologies

- There are several commonly used network topologies, or ways of routing the interconnections.

(i) Star Network:

- This means running a separate cable or line between server and each node. This is useful when a master slave relationship exists between the server and the nodes.

- For stending data and files from one node to another a request should be mode to the server, which establishes a dedicated path between the nodes. The data can be transmitted through this path.

(ii) Ring Network:

- This involves connecting all nodes in series. The cable will normally loop back to form a full circle. This is sometimes used when nodes are widely separated, as each node can act as a repeater for message destined for downstream nodes.

- The data will have to pass through other nodes before reaching the server. The data is sent in the form of a packet which contains both source and destination addresses of the data.

- As the packet circulates through the ring the destination station copies the data into its buffer and the packet continues to circulate until it goes back to source workstation as an acknowledgement.

(iii) Bus Networks:

- This type on interconnection allows all nodes to share the same cable. Any message that travels on the cable is 'seen' by every node on the cable. This topology uses base band and broadband transmission.

3.2 COMPONENT OF NETWOKING

- CIM networking consists of two parts:

    (1)   CIM Network Hardware

(2)　CIM Software

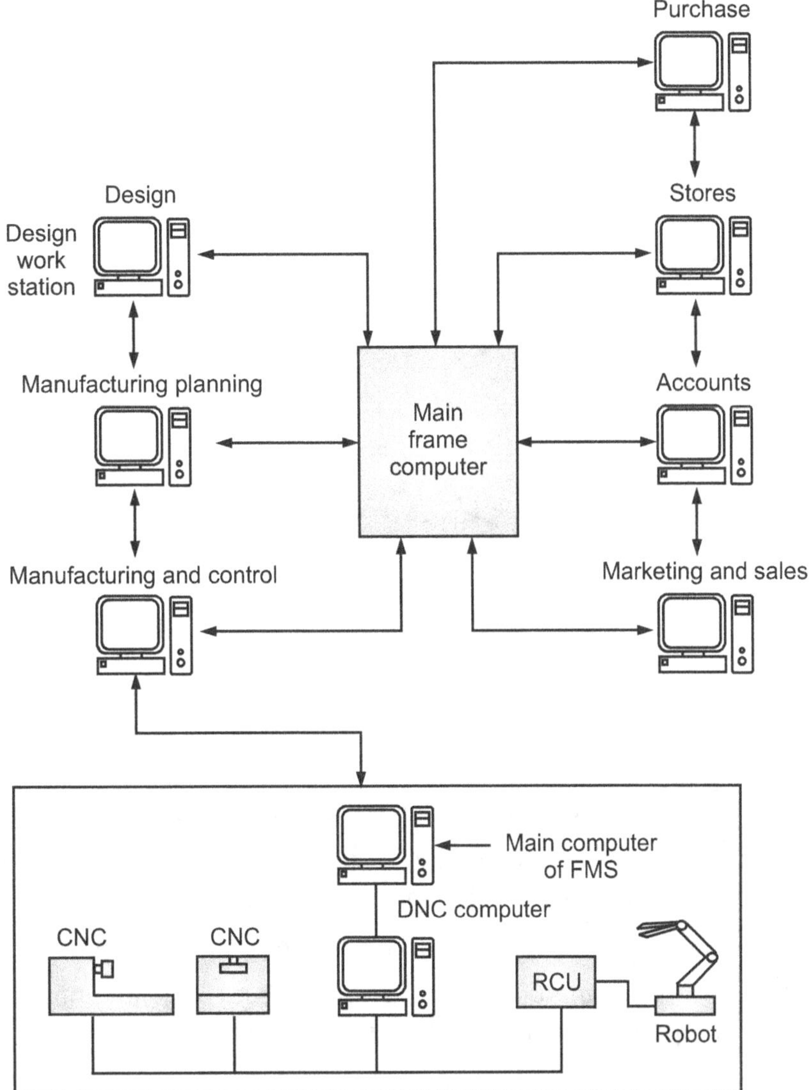

Fig. 3.3 : Typical Computer Hardware Layout of CIM System

3.3 DATA BASE MANAGEMENT SYSTEM (DBMS)

- A database management system (DBMS) defines, creates and maintains a database. The DBMS also allows controlled access to data in the database. A DBMS is a combination of five components hardware, software, users and procedures (Fig. 3.4).

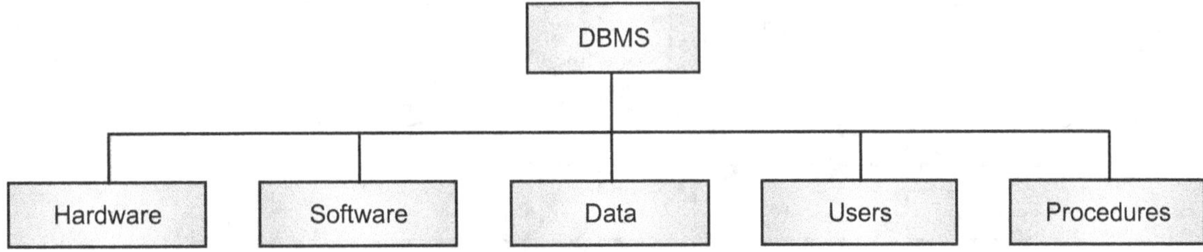

Fig. 3.4 : DBMS Components

- **Hardware:** The hardware is the physical computer system that allows access to data.

- **Software:** The software is the actual program that allows users to access maintain and update data. In addition, the software controls which user can access which parts of the data in the database.

- **Confidentiality:** The data in a database is stored physically on the storage devices. In a database, data is a separate entity from the software that accesses it.

- **Users:** In a DBMS, the term users has a board meaning we can divide users into two categories end users and applications programs.

- **Procedures:** The last component of DBMS is a set of procedure of rules that should be clearly defined and followed by the users of the database.

3.3.1 Data Base Types

- There are three ways in which data can be organized; hierarchical, network or relational.

3.3.1.1 Hierarchical Database

- Fig. 3.5 shows a typical hierarchical file structure. The nodes in level 2 are the children of node at level 1. The nodes at level 2 in turn become parents of nodes in level 3 and so on.

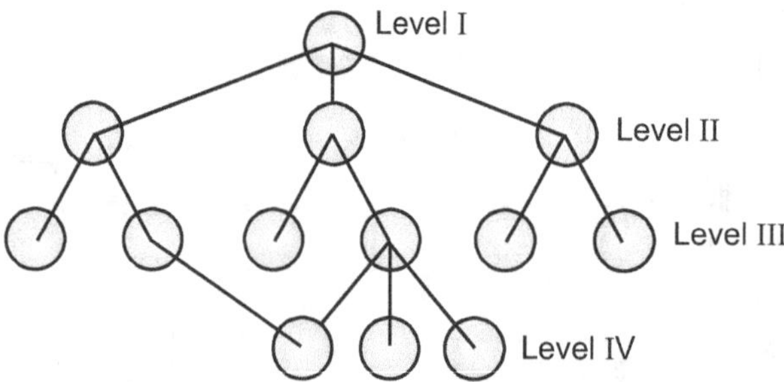

Fig. 3.5 : A Typical Hierarchical File Structure

- In a hierarchical model, data files are arranged in a tree like structure which facilitates searches along branch lines; records are sub-ordinated to other records at a higher level.

- Starting at the root of the tree, each file has a one to many relationship to its branches.

- A parent file can have several children. A good example of such organization might be a parts list in which each product is composed of assemblies which are in turn composed of sub assemblies and/or component parts.

- As an example of hierarchical database structure, the parts list of lathe assembly is shown in Fig. 3.6. Examples of hierarchical database management system are IMS and SYSTEM 2000.

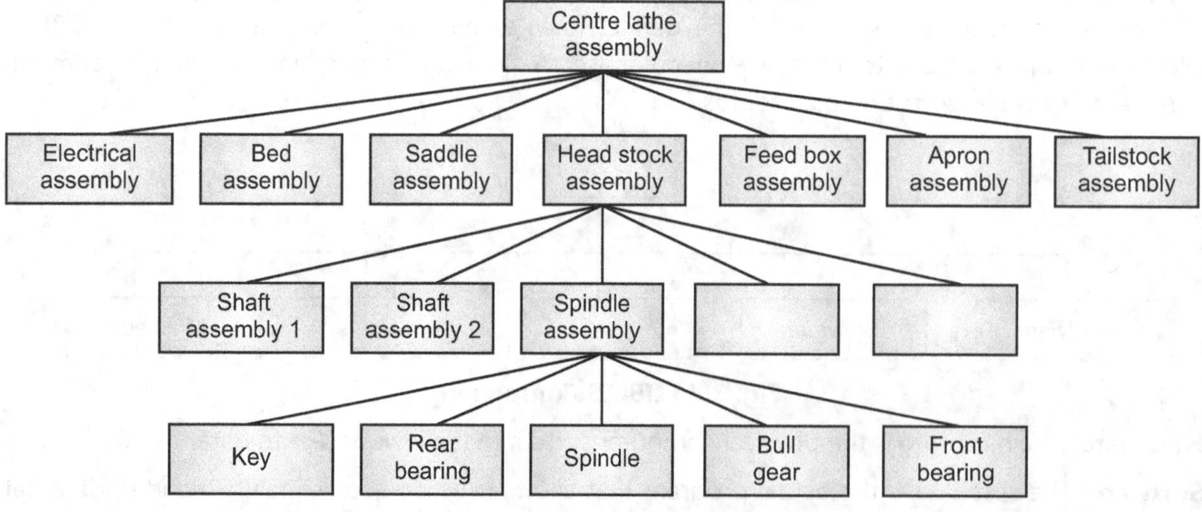

Fig. 3.6 : Parts of a Lathe Assembly

3.3.1.2 Network Data Base

- The network database is a combination of several hierarchies in which child files can have more than one parent file, thereby establishing a many-to-many relationship among data.

- A hierarchical and network databases data relationship are predefined and embedded in the structure of the database. Access to data is processed by associated application programs.

- A limitation of both hierarchical and network systems is the restriction they place on data access. They both require that the rules of data access be defined when the data structure is defined.

- The access rules are difficult to modify after database has been implemented. They are suited for batch operations that are highly structured and repetitive involving high transaction rates.

3.3.1.3 Relational Database Management System

- In a relational database system data is organized in the form of table.

- Relational database connect data in different files by using common data elements or a key field.

- Data in relational database is stored in different tables, each having a key filed that uniquely identifies each row.

- Relational Databases are more flexible than either the hierarchical or network database structures.

- In relational database, tables or files filled with data are called relation, tuples designates a row or record, and columns are referred to as attributes or fields.

3.3.1.4 Object Oriented Data Base

- An object oriented database management system (OODBMS) is a DBMS that facilitates creation and modelling of data as objects. Object-oriented databases use small, reusable pieces of software called objects. The objects themselves are stored in the object-oriented database. Each object consists of two elements:

  (i)   A piece of data (for example, sound, video, text or graphics).

  (ii) Instructions or software programs called methods for what to do with the data.

- This includes support for classes and inheritance of class properties and methods by sub classes and their objects. And OODBMS should satisfy two criteria-it should be a DBMS and it should be object oriented.

- The first criterion involves five features; peristence, secondary storage management, concurrency recovery and query facility.

- The second criterion requires eight features to be incorporated; complex objects, object identity, encapsulation, types or classes, inheritance, overriding combined with late binding, extensibility and computational completeness.

3.3.1.5 Functions of Data Base Management System

A database serves the following functions :

- Reduce or eliminate redundant data.

- Integrate existing data.

- Provide security.

- Share data among users.

- Incorporate changes quickly and effectively.

- Exercise effective control over data.

- Simplify the method of using data.

- Reduce the cost of storage and retrieval of data.
- Improve accuracy and integrity of data.

3.3.1.6 Advantages of DBMS

- A database management system (DBMS) is defined as the software system that allows users to define, create, maintain and control access to the database. DBMS makes it possible for end users to create, read, update and delete data in database. It is a layer between programs and data.

- Compared to the file based Data Management System, Database management system has many advantages. Some of these advantages are given below:

(1) Reducing Data Redundary:

- The file based data management systems contained multiple files that were stored in many different locations in a system or even across multiple systems. Because of this, there were sometimes multiple copies of the same file which lead to data redundancy.

- This is prevented in a database as there is a single database and any change in it is reflected immediately. Because of this, there is no chance of encountering duplicate data.

(2) Sharing of Data:

- In a database, the users of the database can share the data among themselves. There are various levels of authorization to access the data and consequently the data can only be shared based on the correct authorization protocols being followed.

- Many remote users can also access the database simultaneously and share the data between themselves.

(3) Data Security:

- Data security is vital concept in a database. Only authorized users should be allowed to access the database and their identity should be authentically using a username and password.

- Unauthorized users should not be allowed to access the database under any circumstances as it violates the integrity constraints.

(4) Privacy:

- The privacy rule in a database means only the authorized users can access a database according to its privacy constraints.

- There are levels of database access and a user can only view the data he is allowed to. For example, In social networking sites, access constraints are different for different accounts a user many want to access.

(5) Back-up and Recovery:

- Database management system automatically takes care of back-up and recovery. The users do not need to back-up data periodically because this is taken care of by the DBMS. Moreover, it also restores the database after a crash or system failure to its previous condition.

(6) Data Consistency:

- Data consistency is ensured in a database because there is no data redundancy. All appears consistently across the data base and the data is same for all the users viewing the database. Moreover, any changes made to the database are immediately reflected to all the users and there is no data inconsistency.

Important Points

1. **CIM Networking:**
   - Types of network :
     - (i)   LAN (Local Area Network)
     - (ii) WAN (Wide Area Network)
   - **LAN:** Local Area Network (LAN) is privately owned networks covering a small geographic area (less than 1 km), like home, office, building or group of buildings.
   - **WAN:** Wide Area Network is a long-distance communications network that covers a wide geographic area, such as state or country or even the whole world.

2. **Types of Network Topologies:**
   - **Star Network:** This means running a separate cable or line between server and each node. This is useful when a master slave relationship exists between the server and the nodes.
   - **Ring Network:** This involves connecting all nodes in series. The cable will normally loop back to form a full circle. This is sometimes used when nodes are widely separated, as each node can act as a repeats, for message destined for down stream nodes.
   - **Bus Network:** This type on interconnection allows all nodes to share the same cable. Any message that travels on the cable is 'seen' by every node on the cable. This topology use base band and broadband transmission.

3. **Database Management System (DBMS):**
   - A database management system (DBMS) define, creates and maintains a database. The DBMS also allow controlled access to data in the database.
   - A DBMS is a combination of five components:
     - (i)   Hardware
     - (ii)   Software
     - (iii) Users
     - (iv)   Data
     - (v)   Procedures

4. **Database type:**
   - (i)   Hierarchical Database
   - (ii)   Network Database
   - (iii) Relational Database
   - (iv)   Object Oriented Database

5. **Function of Database Management System:**
   - Reduce or eliminate redundant data.
   - Integrate existing data.
   - Provide security.

- Share data among users.
- Simplify the method of using data.
- Reduce the cost of storage and retrieval of data.
- Improve accuracy and integrity of data.

6. Advantage of DBMS:

(i) Reducing Data Redundancy

(ii) Sharing of Data

(iii) Data Security

(iv) Privacy

(v) Backup and Recovery

(vi) Data Consistency

Practice Questions

1. Explain type of networks.
2. Explain characteristics of LAN network.
3. Explain characteristics of WAN network.
4. Type of network topologies.
5. Explain star topology.
6. Explain bus topology
7. Explain ring topology.
8. Explain component of networking
9. Explain type of database.
10. Explain hierarchical database management.
11. Explain rational database of Management.
12. Explain object oriented database.
13. Explain function of database management system.
14. Advantage of DBMS.

GROUP TECHNOLOGY AND FLEXIBLE MANUFACTURING SYSTEM

Weightage of Marks = 12, Teaching Hours = 08

Syllabus

4.1 **Group Technology:** Concept, Basis for developing part families, Part classification and coding with example, concept of cellular manufacturing. Advantages and limitations.

4.2 **Flexible Manufacturing System:** Introduction, Concept, Definition and need, Sub-systems of FMS, Comparing with other manufacturing approaches.

4.3 **Major Elements of FMS:** Workstations, Material handling and storage system, Computer control system and human resources.

4.4 **Classification Based on Flexibility:** Dedicated FMS, Random order.

4.5 **Classification Based on Types of Layouts:** Inline layout type, Rotary layout, Rectangular layout, Loop layout type ladder layout type.

4.6 Applications and benefits of FMS, Advantages and Disadvantages of FMS.

About this Chapter

At the end of this chapter, students will be able to:

- Justify the concept of Group Technology and its benefits for the given situation.
- Classify the FMS based on flexibility for the given types of layouts. .
- Compare the given two manufacturing systems based on the given criteria with examples.
- Justify the use of FMS for the given situation with examples.

4.1 GROUP TECHNOLOGY

4.1.1 Concept

- A batch production is the most common form of production and constitute more than 50% of the total manufacturing activity.
- Therefore, there is a growing need to make batch production more efficient and productive.
- In addition, there is increasing trends towards achieving a higher level of integration between the design and manufacturing activities of a company.
- The above two objectives can be achieved by using a manufacturing philosophy known as Group Technology (GT).
- Group Technology is a manufacturing philosophy in which a similar parts are identified and grouped together as a part family, in order to take the advantage of their similarities in design and manufacturing.

- In manufacturing plant, similar parts are arranged into part families. Each part family, which consists of number of similar parts, possesses similar design and/or manufacturing characteristics.

- For example, a plant manufacturing different types of gear boxes needs to manufacture number of varieties of shafts, gears, keys, spacers, casings etc. By using the philosophy of group technology, these parts are grouped into part families like : shaft, gear, key, spacers and casing.

- The production machines are grouped into machine cells, where each cell specializes in the production of one part family.

4.1.2 Basis for Developing Part Families

Methods of Grouping Parts into Part Families:

- The major obstacle in changing over from conventional process layout to group technology layout is the problem of grouping parts into part families.

- There are four general methods for grouping parts into families:

    (1)   Visual Inspection

    (2)   Composite Part Method

    (3)   Production Flow Analysis (PFA)

    (4)   Parts Classification and Coding.

4.1.2.1 Visual Inspection

- In visual inspection the grouping of the parts into part family is done by looking for similarities in shape, size and methods of manufacture.

- The visual inspection is lest expensive method. However, it is least sophisticated and least accurate method.

- The visual inspection method needs a lot of experience and can only be employed, if number of parts is not very large.

4.1.2.2 Composite Part Method

- In this method, the features of all parts of the part family are combined into a hypothetical composite part as shown in Fig. 4.1.

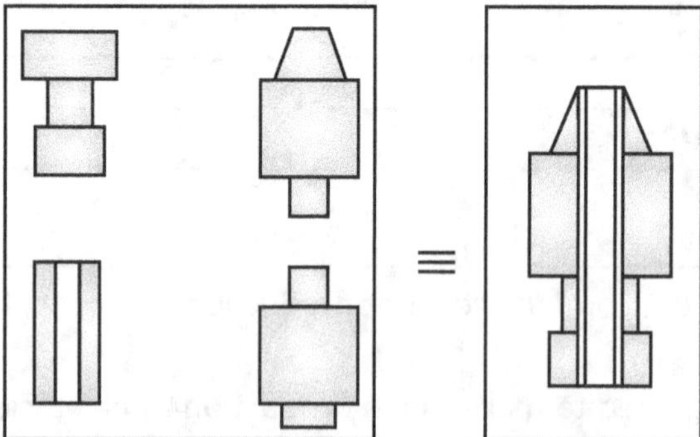

Fig. 4.1: Composite Part Method

- For a hypothetical composite part, the list of all operation is prepared and a tool-setting is done on a multi-tool set-up which can handle all the tools, like turret.

- The list of operations for hypothetical composite part essentially includes all operations required for machining the complete part family.

- Each part of the part family may not need all the operations of hypothetical composite part.

4.1.2.3 Production Flow Analysis (PFA)

- The Production Flow Analysis (PFA) uses manufacturing data to identify part families.

- The production flow analysis involves following steps:

  (1) Data collection: The data such as part number and operation sequence is collected from the manufacturing data contained in the route sheets.

  (2) Sorting of operations: The operations are arranged according to similarly.

  (3) Preparation of PFA chart: The PFA chart containing the data of part numbers against operation or machine code, as shown in Table 4.1 is prepared. The cross mark (⊗) at location of indicates part number 2 requires operation F.

Table 4.1: PFA Chart

Part No. \ Operation or Machine Code	A	B	C	D	E	F	G	H	I	J	K	Remark
1.	⊗	⊗		⊗								Part family - I
2.						⊗	⊗					Part family - II
3.								⊗	⊗		⊗	Part family - III
4.		⊗	⊗									Part family - I
5.								⊗	⊗	⊗	⊗	Part family - III
6.	⊗	⊗		⊗								Part family - I
7.					⊗		⊗					Part family - II
8.										⊗	⊗	Part family - III

- The disadvantage of Production Flow Analysis (PFA) is that, it accepts validity of existing route sheets without checking its consistency or validity.

4.1.2.4 Parts Classification and Coding

- Parts classification is the process of growing of parts of parts on the basis of essential features of the parts, while coding is the process of assigning the codes to the parts.

- In parts classification and coding, the parts are grouped into families by considering:

  (i) Design attributes of each part; and or

  (ii) Manufacturing attributes of each part.

- The attributes of the part are uniquely defined and identified by means of a code number.

- This part classification and coding may be carried out on the entire list of active parts of the film.

- Many parts, classification and coding systems have been developed throughout the world, large number of commercial software are available for part classification and coding.

- However, none of them has been universally adopted. The system best suited for one company may not be suited to another company.

- Activities in parts classifications and coding:

  The parts classification and coding involves two activities :

  (1) Parts classification,   (2) Parts coding.

  These two activities are discussed in subsequent section.

4.1.2.5 Parts Classification Systems

The following three categories of systems are used for parts classification:

(i)   Systems based on part design attributes.

(ii)   Systems based on part manufacturing attributes.

(iii)   Systems based on both design and manufacturing attributes.

(i)   Systems based on part design attributes:

- This category of systems are useful for design standardization:

- The different design attributes used, for parts classification in this type of systems are:

    (a)   Basic external shape.   (b)   Basic internal shape.

    (c)   Length/diameter ratio.   (d)   Material type.

    (e)   Part function.   (f)   Major dimensions.

    (g)   Minor dimensions   (h)   Tolerances.

    (i)   Surface finish.

(ii)   Systems based on manufacturing attributes:

- This category of systems are used for computer aided process planning, tool design and other production related functions.

- The different manufacturing attributes used, for parts classification, in this type of systems are:

    (a)   Major process.   (b)   Minor operations.

    (c)   Major dimension.   (d)   Length/diameter ratio.

    (e)   Surface finish.   (f)   Machine tool.

    (g)   Operation sequences.   (h)   Production time.

    (i)   Batch size.   (j)   Annual production.

    (k)   Fixtures needed   (*l*)   Cutting tools.

(iii) Systems based on both design and manufacturing attributes :

- This category of systems attempt to combine the attributes and advantages of first two types of systems into a single system.

- The different attributes used, for parts classification in this type of systems are:

    (a)   Basic external shape.   (b)   Basic internal shape.

    (c)   Length/diameter ratio.   (d)   Material type.

    (e)   Part function.   (f)   Major dimensions.

    (g)   Minor dimensions.   (h)   Tolerances.

    (i)   Surface finish.   (j)   Major process.

    (k)   Minor operations.   (*l*)   Machine tool.

    (m)   Operation sequence.   (n)   Production time.

- Consider an example of machine tool manufacturing industry. The various part families are:

    (a)   Part family - Heavy parts (beds, columns etc.).

    (b)   Part family - Shafts (large L/D ratios).

    (c)   Part family - Spindles (small L/D ratio).

    (d)   Part family - Non-round parts.

    (e)   Part family - Disc type parts including gears.

4.1.2.6 Parts Coding Systems

- A parts coding systems consists of a sequence of symbols that identify parts design and/or manufacturing attributes.

- The symbols in the code can be : (i) all numeric, (ii) all alphabetic, or (iii) combination of numeric and alphabetic.

- However, most of the standard coding systems use only numbers.

- There are three basic types of parts coding systems used in group technology applications.

  (1) Hierarchical code (monocode):

  - In hierarchical code or monocode, the interpretation of each succeeding digit depends upon the value of preceding digits.

  - For example, consider two parts defined by five digit codes 15689 and 25689. Suppose, first digit stands for the general part shape; 1 means round part and 2 means flat part. If preceded by 1, the 5 may indicate L/D ratio, and if proceed by 2, the 5 may indicate overall length.

  - The hierarchical code provides a relatively compact structure, which conveys a lot of information about the part in a limited number of digits.

  (2) Chain type code (polycode):

  - In chain type code or polycode, the interpretation of each digit in the sequence is fixed and does not depend upon the value of the preceding digits.

  - The advantages of chain type code is, each digit has fixed meaning and is independent of other digits. Therefore, the coding system is more universal in nature.

  - However, the chain type codes tend to be relatively long.

  (3) Hybrid code:

  - The hybrid code is a combination of hierarchical and chain type structures. It is an attempt to achieve the best features of monocodes and polycodes.

  - Most of the commercial parts coding systems used in industry are hybrids codes.

4.1.2.7 Commercial Parts Classification and Coding Systems

- The large number of parts classification and codes systems are commercially available.

- Some of the important systems are listed below:

  (a) OPTIZ system. (b) CODE system.

  (c) BRISCH system. (d) KK-3 system.

  (e) MICLASS system. (f) DCLASS system.

  (g) COFORM system. (h) TOSHIMA system.

- The OPTIZ part classification and coding system is discussed in next section.

4.1.2.8 OPTIZ Part Classification and Coding System

- OPTIZ part classification and coding system is the most widely used and perhaps the best classification and coding system available today.

- The basic code consists of nine digits, which can be extended by adding four more digits, as shown in Fig. 4.2.

- **Form code:** The first five digits are called 'form code' and describe the primary design attributes of the part.

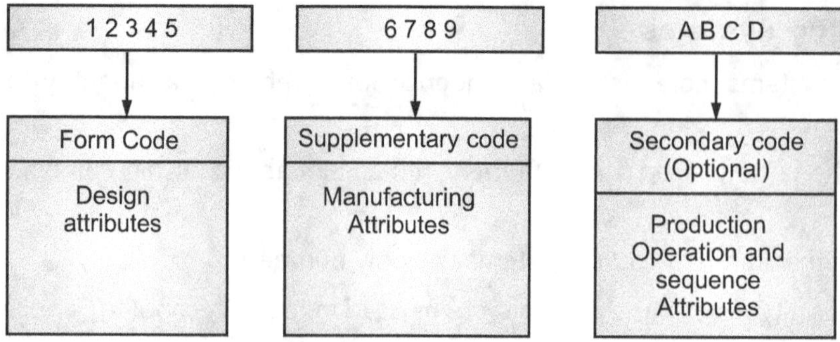

Fig. 4.2: Basic Structure of OPTIZ Code

- **Supplementary code:** The next four digits are called 'supplementary code' and describe the manufacturing attributes of the part.

- **Secondary code:** The extra four digits are called 'secondary code' and are intended to identify the production operation type and sequence. The secondary code can be developed by the individual industry as per its own requirements. Fig. 4.3 shows the complete structure of OPTIZ code.

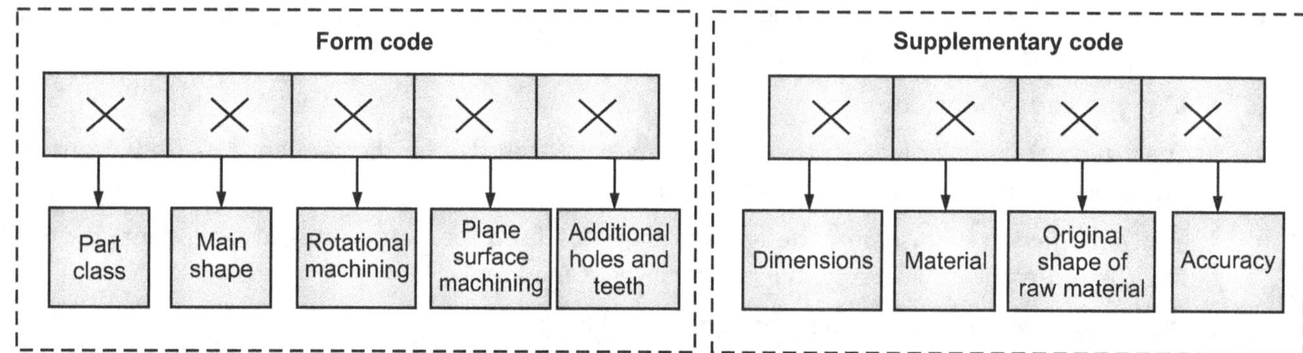

Fig. 4.3: Complete Structure of OPTIZ Code

4.1.3 Concept of Cellular Manufacturing

- In group technology, the production machines are grouped into machine cells, where each machine cell specializes in the production of one part family.

- **Cellular manufacturing:** The group technology philosophy of grouping of production machines into machine calls, where each machine cell specialize in the production of one part family is called cellular manufacturing.

4.1.3.1 Objective of Cellular Manufacturing

The basic objectives of cellular manufacturing are:

(i)   To reduce work-in-process inventory.

(ii) To reduce manufacturing lead time.

(iii) To reduce work part handling.

(iv) To simplify production scheduling.

(v)   To reduce set-up times, and

(vi) To reduce process variation.

4.1.4 Advantages of Group Technology

The group technology offers the following advantages:

1. **Reduced material handling:** A group technology layout is designed on the basis of minimizing the material flow. As the parts are moved within a machine. Cell rather than within the entire factory, the material handling is reduced.

2. **Reduced tool set-up time:** The group technology reduces tool set-up time, thereby reducing the manufacturing lead time.

3. **Reduced work-in-process:** The group technology drastically reduces the manufacturing lead time. This reduces the work-in-process and hence, leads to the reduction in inventory of raw materials.

4. **Promotes standardization of tooling, fixture and set-ups:** Group technology manufacturers the similar parts in one cell. This leads to standardization of tooling and fixtures.

5. **Simplified process planning and production scheduling:** In group technology, the parts are manufactured in a machine cell with simplified material flow. Therefore, process planning and production scheduling get simplified.

6. **Better worker satisfaction:** In group technology, the quality of part is attributed to a group of workers in a machine cell. Therefore, the workers feel work responsible and motivated for the parts leaving from their machine cell.

7. **Better product quality and productivity:** The above mentioned features of group technology lead to improvement in product quality as well as productivity.

Limitations of Group Technology:

The limitation of group technology are as follows:

1. **Difficulty in grouping the parts into families:** In a factory manufacturing large number of parts, grouping of parts into families is a difficult task and consumes significant time.

2. **Difficult in rearranging machines into machine cells:** It is time consuming and costly to rearrange the machines into machine cells.

3. **Inertia to change:** Normally there is a resistance, from worker, for any change in manufacturing system.

4.2 FLEXIBLE MANUFACTURING SYSTEM

4.2.1 Introduction

- **Flexible Manufacturing System (FMS)** is a highly automated group technology machine cells, consisting of group of workstations (CNC macyines or CNC machining centers), interconnected by an automated material handling and storage systems and controlled by a computer system.

- FMS is a capable of processing variety of parts. The system set-up and processes are programmable and can be programmed as per the requirement of a part.

- The basic features of flexible manufacturing systems (FMS) are:

  (i) Ability to manufacture variety of products.

  (ii) Less manufacturing lead time.

  (iii) High quality; and

  (iv) Low cost.

4.2.2 Definition and Concept

- **Definition:**

  Flexible Manufacturing System (FMS) is a computer controlled process technology suitable for producing a moderate variety of products in moderate, flexible volumes.

- **Concept:**

  Before we proceed to understand the meaning of a flexible manufacturing system (FMS) first lets revise the individual meaning of three important words that makes a term called FMS.

In context of this article, individual meaning of three important words viz., flexible, manufacturing and systems are stated below:

1. **Flexible** means the capability (potential) of any process (system) whether automated or manual to resume its original form after performing action or series of actions, such as strengthening, compression, melting, recycling etc.

2. **Manufacturing** includes any of those processes (manual or automated) which are primarily or supplementary required for the production of a marketable finished product (goods).

3. **System** can be explained as a logical and functional arrangement of interrelated activities that operate altogether to achieve a common goal.

With this basis revision, now let's move ahead and discuss the meaning of a flexible manufacturing system.

Flexible manufacturing system is abbreviated as FMS. FMS is a computer-controlled system. It is an automated factory. It contains many workstations. Each workstation contain a set of computer-controlled machines. Each workstation performs a different function (job).

For example, One workstation does only drilling of holes, another workstation focuses solely on painting etc. The full FMS is controlled by a central computer. Here, materials are moved from one workstation to another and form one machine to another by computer-controlled machines and robots.

So, the full production process is dividend, into many workstations. Each workstation performs one specific function. The finished goods of one workstation is the raw material of the next workstation and so on. The materials are moved by automatic machines. The materials are loaded and unloaded on the machines by robots.

4.2.3 Need of Flexible Manufacturing System

- In a present competitive market, the customer is a center of focus. In order to maintain and increase the market share, the manufacturing industries are compelled to:
  - (i)   innovate and keep on modifying the product;
  - (ii)   reduce the delivery time; and
  - (iii) quote competitive prices even for small orders.
- Therefore, the manufacturing facility has to meet the following requirements:
  - (i)   Flexible manufacturing system to incorporate the product changes at short notice to meet customer's requirement.
  - (ii)   Automation of manufacturing system so as to integrate all functions of manufacturing.
  - (iii) Reduction in lead time.
  - (iv)   High productivity for all batch sizes, large or small.
  - (v)   Reduction in material handling.
  - (vi)   Reduction in material handling.
- The use of stand alone CNC machine tools gives the flexibility to manufacturing system but does not provide automation and integration of all functions of manufacturing.
- The mass production manufacturing facilities such as transfer lines provide the automation and integration to manufacturing but lacks in flexibility.
- In short, neither of these two options, fulfill the basic requirements of the modern manufacturing facility.
- This has created a need to evolve the alternative manufacturing system which:
  - (i)   Provides automation to manufacturing system and integrates all its functions; and
  - (ii)   Given flexibility to manufacturing system to incorporate the product changes at short notice.
- Such manufacturing system is known as flexible manufacturing system and it meets all the six requirements listed above.

4.2.4 Subsystems of FMS

There are three major subsystems in FMS:

(i) Computer-controlled manufacturing equipment (for example, numerically controlled machine tools, robots, gantry loaders, palletizing systems, washing stations, tool pre-setters, in-process inspection systems etc.

(ii) Automated materials storage retrieval, transport and transfer system.

(iii) Manufacturing control system (includes both machine tool, tool and logistics control.

Some FMS's have additional sub-systems. For example, in a machining application there may also be systems for presetting tools, storing and retrieving tools, disposing of chips and cutting fluids, washing and inspection workpieces. These subsystems must be linked together to achieve integrated manufacturing operations.

4.2.5 FMS Comparing with Other Manufacturing Approaches

On-Off and low volumes of production are normally carried out by conventional general purpose machine tools. When the number of parts in a production run is more it is called batch production. A batch production shop is best suited for small quantities of many different types of parts. The very nature of production makes the operation of a job shop less efficient than an automated production line.

Since, the job shop must be provided the greatest degree of flexibility, most of its operations are manual. They are normally equipped with general purpose CNC machine tools. Hard automation with dedicated equipment is best suited for the production of very large quantities of identical parts. Production of automobile components in a transfer line falls under this category. A large portion of the manufacturing industry involves the intermediate level of batch operations that lend themselves to the FMS approach. In this case, volume is less but varieties is less but varieties are more.

FMS thus basically attempts to efficiency automate batch manufacturing operations. They are an alternative that fits in between the manual job shop and hard automation. FMS is best suited for applications that involve an intermediate level of flexibilty and low or medium quantities. Fig. 4.4 shows the different types of production systems and it can be seen from the figure that FMS fits into the intermediate range of production. General purpose machines can accommodate a large variety of parts. They are manually operate and therefore production volumes are low. CNC machines can accommodate variety but the production volume is less as the machines are not optimized for the highest productivity for a specified type of a job. It can be seen that FMC and FMS satisfy both variety and volume equally well. If we take special purpose machines. Variety is must restricted. Transfer lines are dedicated usually to manufacture a component and hence can be said to have the minimum variety.

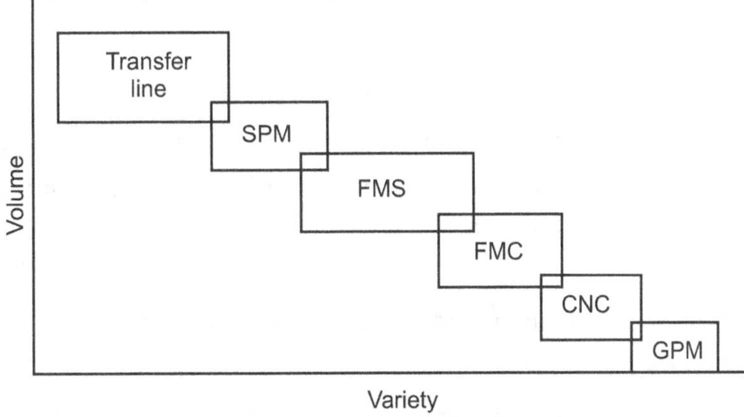

Fig. 4.4: Types of Production Systems

4.3 MAJOR ELEMENTS OF FMS

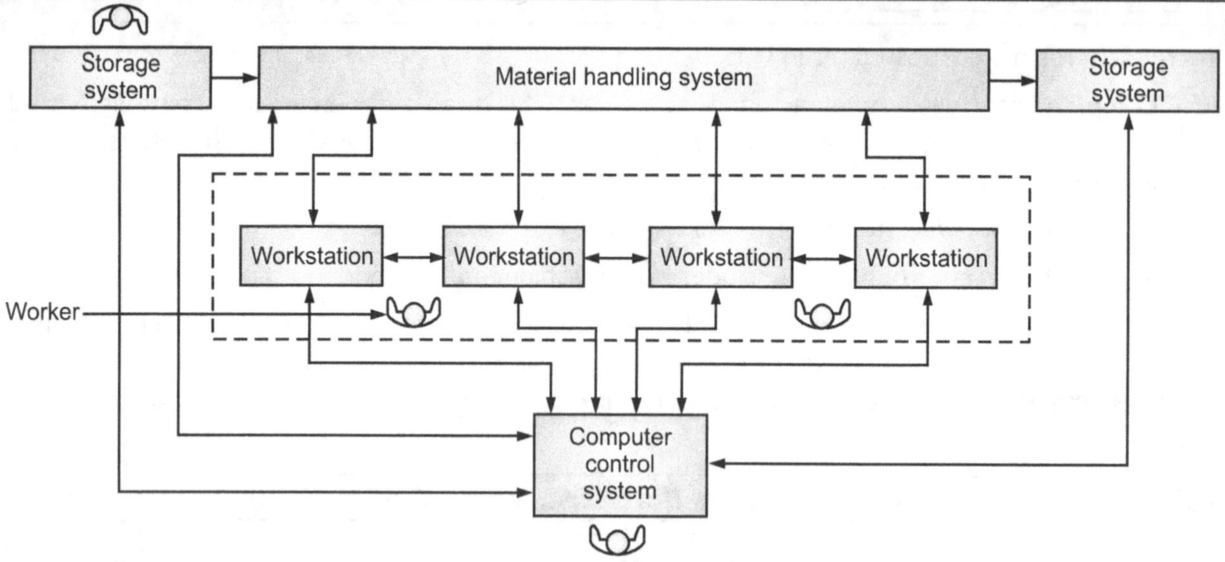

Fig. 4.5: Elements of Flexible Manufacturing System

Any flexible manufacturing system consists of four basic elements [Fig. 4.5]

(1) Workstations:

- The main element of any flexible manufacturing system is workstations.

- The flexible manufacturing system consists of one or more workstations which include: CNC machines, CNC machining centers, CNC presses, CNC forging presses, industrial robots, heating furnaces etc.

- The type of workstations used and their sequences depend upon the part to be manufactured.

(2) Material Handling and Storage System:

- The second major element of flexible manufacturing system is material handling and storage system.

- The material handling system performs the following functions:

    (i)   Sequential or random movement of workpieces between workstaions.

    (ii)   Handling and locating the workpieces of different configurations.

    (iii) Temporary storage of workpieces, and

    (iv)   Loading and unloading of workpieces.

- The storage system performs the functions of storage of raw material and storage of finished parts.

(3) Computer Control System:

- The computer control system is the brain of flexible manufacturing system.

- It is interfaced with workstations, material handling system, and storage system.

- The computer control system integrates, monitors and controls the functioning of workstations, material handling system and storage system.

(4) Human Resources:

- Human resource is needed to manage the operations of flexible manufacturing system.

- The functions performed by operator include:

    (i)   Loading of raw material into the system.

(ii)   Unloading of finished parts from the system.

(iii) Changing and setting of tools.

(iv) Programming and operating the systems.

(v)   Maintenance of system, and

(vi) Overall management of system.

4.4 CLASSIFICATION BASED ON FLEXIBILITY

Based on the flexibility of system, the flexible manufacturing systems can be classified into two types:

(1)   Dedicated or special FMS.

(2)   Random-Order FMS.

(1) Dedicated or Special FMS is one which is designed to produce a limited variety of part configuration.

- The design of the parts to be manufactured by the system is known in advance i.e. before designing to system. Therefore, the system can be designed with a certain amount of process specialization to make machining operations more efficient.

- Instead of using general purpose machines, the special purpose machines suitable for limited part family are used. This increase the rate of production of the system.

(2) Random-Order FMS:

- Random-order FMS is used when the part family is large and there are substantial variations in part configurations.

- In order to accommodate these variations in part configurations, the random-order FMS uses general purpose machines.

- The use of general purpose machines adds flexibility to the system. However, this reduces the rate of production of the system.

4.5 CLASSIFICATION BASED ON TYPES OF LAYOUTS

The different layouts used in the flexible manufacturing systems are as follows:

(1) Inline layout type FMS.

(2) Rotary layout type FMS.

(3) Loop layout type FMS.

(4) Rectangular layout type FMS.

(5) Ladder layout type FMS.

4.5.1 Inline Layout Type FMS

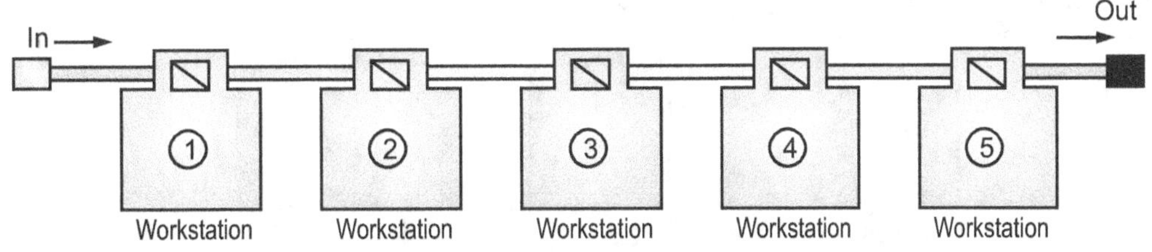

Fig. 4.6: Inline Layout Type FMS

- In an inline layout type FMS, the workstations are arranged in a straight line as shown in Fig. 4.6.

- The parts flow only in one direction and that too in a straight line.

- It is the simplest form of layout and simplifies the material handling system.

4.5.2 Rotary Layout Type FMS

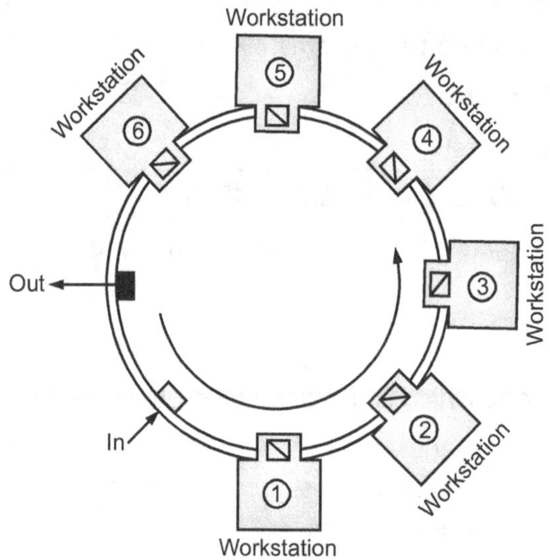

Fig. 4.7: Rotary Layout Type FMS

- In a rotary layout type FMS, the workstations are arranged in a circular arrangement as shown in Fig. 4.7.

- The rotary layout type arrangement is compact and it also simplifies the material handling system.

4.5.3 Loop Layout Type FMS

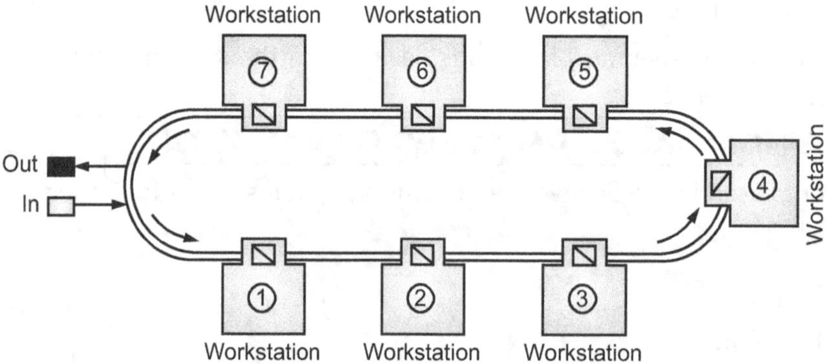

Fig. 4.8: Loop Layout Type FMS

- In a loop layout type FMS, the workstations are arranged in a loop as shown in Fig. 4.8.

- In this layout, the loading and uploading stations are located at one end of the loop.

4.5.4 Rectangular Layout Type FMS

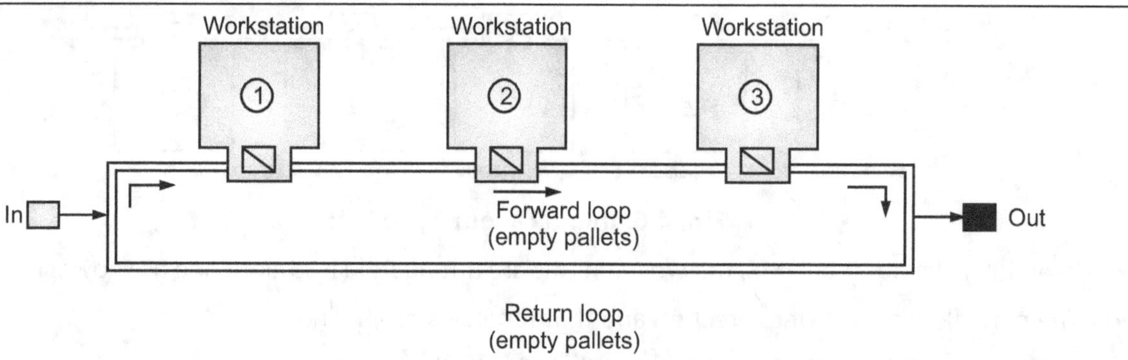

Fig. 4.9: Rectangular Layout Type FMS

- In a rectangular layout type FMS, the workstations are arranged as shown in Fig. 4.9.
- This arrangement is a modification of inline layout. This arrangement is used to return the pallets to the starting position.

4.5.5 Ladder Layout Type FMS

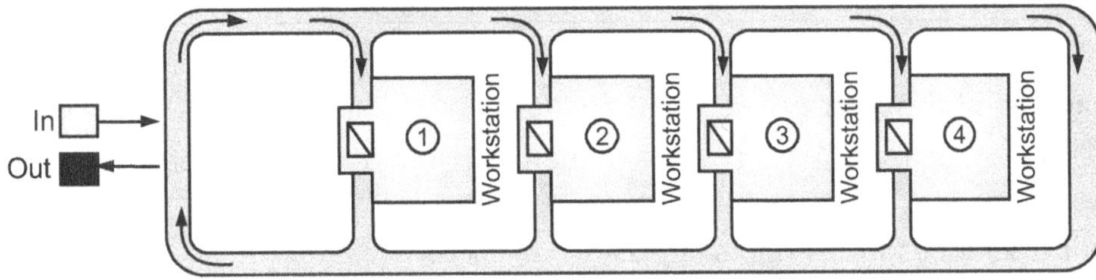

Fig. 4.10: Ladder Layout Type FMS

- In a ladder layout type FMS, the workstations are arranged in the form of rungs of a ladder as shown in Fig. 4.10.
- The rungs increase the possible ways of getting from one machine to the next. This reduces the transport time between workstations.

4.6 APPLICATION AND BENEFITS OF FMS, ADVANTAGES AND DISADVANTAGES OF FMS

4.6.1 Applications and Benefits of FMS

The four types of production systems can be distinguished in term of production quantity (volume) and product variety (flexibility) as shown in Fig. 4.11.

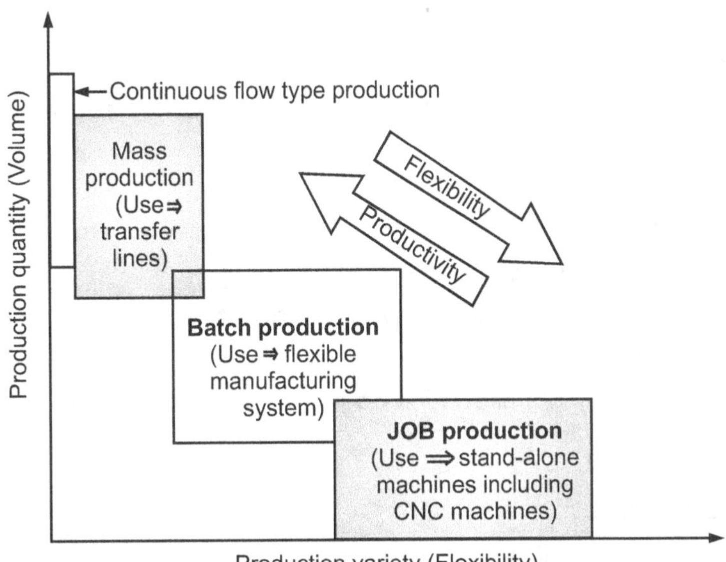

Fig. 4.11: Production Quantity and Product Variety for Production System

- **The stand alone machines including CNC machines** are highly flexible and most suited for large product varieties. However, they are not suitable for large production quantity.
- **The transfer line** are highly efficient in producing large quantities of a fixed product. However, they are highly inflexible and cannot handle large product variety.
- There is a large gap between the areas of functioning of **stand-alone machines** and **transfer lines.** This gap is bridged by flexible manufacturing system.

- The **flexible manufacturing system** is most suitable for products of medium quantity and medium variety.

- For industries dealing with the products of medium quantity and medium variety, the flexible manufacturing system is the solution.

4.6.2 Advantages of Flexible Manufacturing Systems

The flexible manufacturing systems offer the following advantages:

(1) Increased machine utilization:

FMS achieve higher utilization of machines than conventional batch production systems due to following reasons:

- Continuous operations for 24 hours per day,
- automatic tool changing,
- automatic loading and unloading, and
- automatic material handling.

(2) Lesser number of machines required:

Due to higher machines utilization, lesser number of machines are required in FMS.

(3) Reduction in floor space requirement:

As compared to conventional batch production layout, FMS requires lesser floor area.

(4) Greater flexibility:

FMS is more flexible and responsive to change in product design, production schedule, and batch size.

(5) Reduction in inventory:

As different parts are processed together rather than separately in batches, the work-in-process (WIP) is less than that in the conventional batch production system. Therefore, in FMS, the inventory requirement is less.

(6) Lower manufacturing lead time:

Reduced work-in-progress (WIP) time reduces manufacturing lead time. Therefore, with FMS, customer deliveries are faster.

(7) Reduced labour requirement:

In FMS the reliance on labour is less. Hence, it reduces the labour requirement.

(8) Better product quality and productivity:

The FMS results in improved product quality as well as productivity.

(9) Extended period production:

The high level of automation in FMS allows it to operate for extended periods of time without human attention.

4.6.3 Disadvantages of FMS

The disadvantages of FMS are as follows:

(1) High initial cost:

FMS requires high initial investment.

(2) High maintenances cost:

In FMS, the equipment/machines are highly specialized in nature. Hence, highly skilled manpower is required to maintenance cost.

(3) Worker unemployment:

FMS reduces manpower requirement. This leads to unemployment problems.

Important Points

1. Group Technology:

- Group Technology is a manufacturing philosophy in which a similar parts are identified and grouped together as a part family, in order to take the advantages at their similarities in design and manufacturing.

2. Basis for developing part families:

- There are four general methods for grouping parts into families:

  (1) Visual Inspection

  (2) Composite Part Method

  (3) Production Flow Analysis (PFA)

  (4) Parts Classification and Coding

3. Cellular manufacturing:

- The group technology philosophy of grouping of production machines into machine cells, where each machine cell specialize in the production of one part family is called cellular manufacturing.

4. (A) Advantage of group technology:

- Reduce material handling.
- Reduce tool set-up time.
- Reduce work-in-process.
- Promotes standardization of tooling, fixture and set-ups.
- Simplified process planning and production scheduling.
- Better worker satisfaction.
- Better product quality and productivity.

(B) Limitation of group technology:

- Difficulty in grouping the parts into families.
- Difficult in rearranging machine into machine cells.
- Inertia to change.

5. Flexible Manufacturing System (FMS):

- Flexible Manufacturing System (FMS) is a computer controlled process technology suitable for producing a moderate variety of products in moderate, flexible volumes.

(A) Need of flexible manufacturing systems:

- Reduction in lead time.
- High productivity.
- Reduction in manpower.
- Reduction in material handling.

(B) Elements of flexible manufacturing systems:

- Workstation.
- Material handling and storage system.
- Computer control system.
- Human resources.

6. (A) Classification based on flexibility:

- Dedicated or special FMS.
- Random-order FMS.

(B) Classification based on types of layouts:

- Inline layout type FMS.
- Rotary layout type FMS.
- Loop layout type FMS.
- Rectangular layout type FMS.
- Ladder layout type FMS.

7. (A) Advantage of flexible manufacturing systems:

- Increased machine utilization.
- Lesser number of machine required.
- Reduction in floor space requirement.
- Greater flexibility.
- Reduction in inventory.
- Lower manufacturing lead time.
- Reduce labour requirement.
- Better product quality and productivity.
- Extended period production.

(B) Disadvantage of flexible manufacturing system:

- High initial cost.
- High maintenance cost.
- Worker unemployment.

Practice Questions

1. Short note on group technology.
2. Explain part classification and coding with example in group technology.
3. Explain concept of cellular manufacturing.
4. Advantage and limitation of group technology.
5. Short note on flexible manufacturing system.
6. Need of flexible manufacturing system.
7. Short note on subsystem of FMS.
8. FMS comparing with other manufacturing approaches.
9. Explain major elements of FMS.
10. Explain classification based on flexibility dedicated FMS and random order.
11. Explain classification based on type of layout:
    (a) Inline layout type
    (b) Rotary layout type
    (c) Rectangular layout type and ladder type.
12. Explain application and benefit of FMS.
13. Advantage and disadvantage of FMS.

☝ ☝ ☝

Chapter 5

AUTOMATON

Weightage of Marks = 16, Teaching Hours = 12

Syllabus

5.1 **Automation:** Define, Need of automation, High and low cost automation examples of automations.

5.2 **Elements of automation:** Power source, Control unit and feedback control.

5.3 **Types of automation:** Fixed (Hard) automation, Programmable automation and Flexible automation (Soft), Comparison of types of automation.

5.4 **Strategies in automation:** Simplification specializations of operations, multiple operations, Integration of work stations, Increased flexibility, Automated material handling storage system, Online inspection, Online monitoring, Processes control and optimization, Control of plant operations and computer integrated manufacturing.

About this Chapter

At the end of this chapter, students will be able to:

- Explain the main elements of the given automation system.
- Explain the given types of automations with respect to their characteristics.
- Justify the need of automation for the given situation.
- Explain the kind of strategies to be considered while designing automation in industry for the given situation.

5.1 AUTOMATION

5.1.1 Define

- **Automation** can be defined as the technology used for the application of integrated mechanical, electronic and computer based systems in the operation and control of production systems.
- Automation of production system means:
  - (i)   Automation of manufacturing facilities, or
  - (ii) Automation of manufacturing support systems, or
  - (iii) Automation of both the facilities and the manufacturing support systems.
- Automation is not new but has been in use in the industry since quite sometime. The automation seems to be a feasible solution for improving the productivity, quality and economy.
- Automation is not economically feasible in low scale production. However, as the volume of production goes on increasing, automation becomes more and more economically feasible.

5.1.2 Need for Automation

(1) To increase productivity:

- The automation of manufacturing operations usually increases production rate. This means greater output per hour of labour input.
- Thus, automation leads to increase in labour productivity.

(2) To reduce cost of production:

- The automation reduces the labour cost and increases the rate of production, thereby reducing the cost of production.

(3) To improve product quality:

- The automation not only results in higher production rates than manual operations but also improves the product quality.

(4) To mitigate the effects of labour shortages:

- In developed countries where there is shortage of labour, automated operations are used as substitute for labour.

(5) To reduce production time:

- The automation reduces the time required for manufacturing the product.

(6) To avoid high cost of not automating:

- The automation exhibits overall benefits like: improved product quality, high rate of production, higher salaries, better labour relations, better customer satisfaction and better company image.
- The companies without automation are likely to find themselves in a disadvantageous position as compared to the companies with automation.

(7) To have better control over manufacturing activities:

- Automation provides better control over entire manufacturing activity of a company.

(8) To improve worker safety:

- The automation can completely replace the worker, especially in hazardous operations like: spray paintings, welding, chemical processing etc., thereby improving the worker safety.
- The automation has changed the role of worker from active participation to a supervision.

(9) To reduce or eliminate routine manual and clerical tasks:

- Automation reduces/eliminates routine manual and clerical tasks which are boring, fatiguing and isksome, thereby improving the general level of working conditions.

5.1.3 High and Low Cost Automation

High Cost Automation:

High cost automation involves of manufacturing facilities or/and manufacturing support system. If replacement of conventional machine by CNC machines, material handling by conveyors and AGV's.

High cost automation demands huge capital investment and only scale industries can altered it.

Example of high cost automation, full automation factory.

Low Cost Automation:

Low cost automation is the introduction of simple pneumatic, hydraulic, mechanical and electrical devices into the existing production machinery, with a view of improving their productivity.

India is fast developing industrially. The industrial growth, particularly during the last decade has been considerable. This growth of course is not free from the attendant problems. In a country like ours, which has mainly an agricultural based economy, the industrial development is very likely to change the existing conditions.

Low cost automation should not be regarded in terms of a specified maximum capital outlay, but as an approach to automation using equipment and control devices that are, in general, both technically and economically, within the scope of the company concerned.

The main aim of low cost automation is to increase productivity and quality of products and reduce the cost of production and not reduce labour.

5.1.4 Example of Automation

Some of the examples of automation of process in production systems are as follows:

(i)   NC and CNC machines.

(ii)   Automatic assembly machines.

(iii)   Automated transfer lines.

(iv)   Automatic assembly lines.

(v)   Industrial robots.

(vi)   Automated material handling systems.

(vii)   Automated storage systems.

(viii) Automated inspection and quality control systems.

(ix)   Automated feedback and process control equipment.

(x)   Computer aided production, planning and control.

5.2 ELEMENTS OF AUTOMATION

(1) Power Source:

An automated system is used to operate some process and power is required to drive the process as well as controls. There are many source of power available, but the most commonly used power is electricity. The actions performed by automated systems are generally of two types.

(a)   Processing.

(b)   Transfer and positioning.

In the first case, energy is applied to accomplishe some processing operation on some entity. The process may involves shaping, moulding or loading and unloading. All these actions need power to transfer the entity from one state or condition into more valuable state or condition.

The second type of actions - transfer and positioning. In these cases, the product must generally be moved from one location to another during the series of processing steps.

(2) Program of Instructions:

The actions performed by an automated process are defined by a set of instructions known as process. The programmed instructions determine the set of actions that is to be done automatically by the system. The program specifies what automated system should do and how its various components must function in order to accomplish the desired results.

(3) Control System:

The control element of the automated system executes the program of instructions.

The controls in an automated system can be:

(a)   Closed loop.   (b)   Open loop.

(a)   Closed loop control system: It is also known as a feedback control system. In this system the output variable is compared with an input parameter and any difference between the two is used to drive the output into agreement with input.

1.   **Input parameter:** As set point, represents the desired value of output.

2.   **Output variables:** Actual value of parameter.

3.   **Sensors:** A sensor is used to measure the output variable and close the loop.

4.   **Between input and output:** It performs feedback functions.

5.   **Controller:** The controller compares the output with the input and makes the required adjustment in the process to reduce the difference between them.

6. **Actuator:** The adjustment being done with one or more actuators which are the hardware devices that physically carry out the control actions such as, electric motor, cylinder etc.

(b) Open loop control system: It is without the feedback loop. In this case the controls operates without measuring the output variables, so no comparison is made between the actual value of the output and desired input parameters. There is always risk that the actuator will not have intended effect on the process.

5.3 TYPES OF AUTOMATION

The automation of production systems can be broadly classified into three types, as shown in Fig. 5.1.

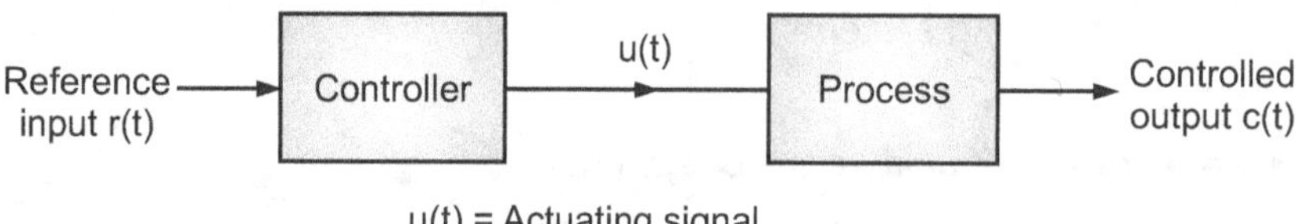

Fig. 5.1: Types of Automation

5.3.1 Fixed (Hard) Automation

- **Fixed (Hard) automation** is an automaton system in which the sequence of processing operations is fixed by the production equipment (machines) configuration. Each processing operation in a sequence is simple.

- The fixed automaton can not be changed once it is established and hence, it is inflexible in accommodating the product variety.

- The fixed automation is economical when there is a continuous high demands for the product at the high volume.

- The initial cost of the automated equipment can be spread over a very large number of units, thus, making the unit cost attractive compared with the equipment without automation.

- The fixed automation is suitable for continuous flow type production systems and mass production system.

- **Features of fixed (hard) automation:**
  (i) High initial investment for custom - engineered equipment.
  (ii) High production rates.
  (iii) Highly inflexible in accommodating product variety.
  (iv) Suitable for continuous flow type production systems and mass production systems.
  (v) No tool set up time, as tooling is fixed.

- **Examples of fixed automation:** Bottling plants, packaging plants, transfer lines etc.

5.3.2 Programmable Automation

- **Programmable automation** is an automation system in which the production equipment (machines) are designed with a capability to change the sequence of operations so as to accommodate the different product configurations.

- The operations sequences is controlled by a program, which is a set of coded instructions that can be read by the equipment.

- New programs can be prepared and entered into the equipment to produce the new products.

- To produce a batch of new product, a new program must be prepared and entered into the equipment (machine). The physical setup of the machine (i.e. tooling, fixtures, machine settings etc.) must also be changed. this change over procedure takes time and is called as set up time.

- Programmable automation is suitable for batch production systems.

- **Features of programmable automation:**
  - (i) High initial investment in general purpose equipments.
  - (ii) Lower production rates than fixed automation.
  - (iii) Flexible in accommodating product variety.
  - (iv) Most suitable for batch production systems.
  - (v) Tool setup time varies from batch to batch.
- **Examples of programmable automation:** NC machines tools, industrial robots, programmable logic controllers etc.

5.3.3 Flexible (Soft) Automation

- **Flexible (soft) automation**, which is an extension of programmable automation is an automation system capable of producing products of design variations, continuously with virtually little or not time loss for changeovers from one product to the other.
- There is virtually no production time loss while reprogramming the system for new configuration of product. Therefore, the system can produce various combinations of products continuously instead of requiring that they be made in batches.
- However, it is important to note that the variety of products that can be produced by flexible automation system is less than that can be produced by programmable automation system.
- **Features of flexible (soft) automation:**
  - (i) High initial investment for custom-engineered equipment.
  - (ii) Medium production rates.
  - (iii) Flexible in accommodating product design variations.
  - (iv) Suitable for continuous production of variable products.
  - (v) Minimal tool setup time.

5.3.4 Comparison of Types of Automation Systems

- Comparison of three types of automation systems is summarized in Table 5.1.

Table 5.1: Comparison of three types of automation systems

Comparison parameters	Fixed (Hard) Automation	Programmable Automation	Flexible (Soft) Automation
Initial Investment	High initial investment for custom-engineered equipment.	High initial investment for general purpose equipment.	High initial investment for custom-engineered equipment.
Production rates	High.	Low to medium.	Medium.
Flexibility	High inflexible.	Flexible in accommodating changes in product variety.	Flexible in accommodating product design variations.
Production system	Suitable for continuous flow type production system and mass production systems.	Suitable for batch production systems.	Suitable for continuous production of variable products.
Tool setup time	No tool setup time as tooling is fixed.	Tool setup time varies from batch to batch.	Minimal tool setup time.

- The three types of automation systems can be distinguished in terms of production quantity (volume) and product variety (flexibility) as shown in Fig. 5.2.

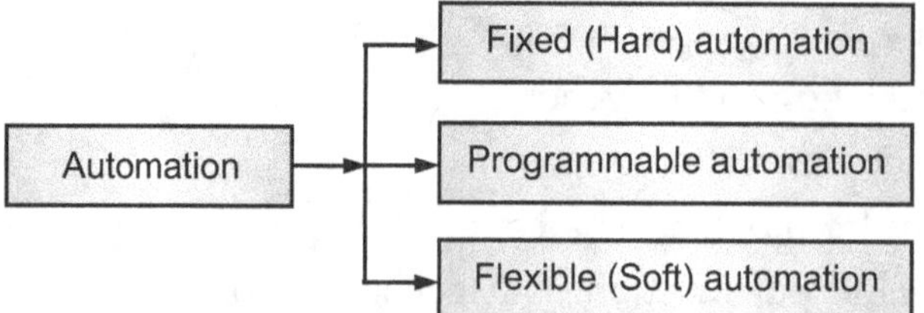

Fig. 5.2: Production quantity and product variety for different automation system

5.4 STRATEGIES IN AUTOMATION

There are certain fundamental strategies that can be employed to improve productivity in manufacturing operations technology. These are referred as automation strategies.

5.4.1 Simplification

The main strategy is simplification process in automation, with the help of this strategy, we done reduces wasted time to do routine processes. Improves quality by requiring accurate information at every step and save money also, because of significant reduced numbers and associated rework.

5.4.2 Specialization of Operations

The first strategy involves the use of special purpose equipment designed to performed one operation with the greatest possible efficient. This is analogous to the concept of labour specializations, which has been employed to improve labour productivity.

5.4.3 Multiple Operation (Combined Operation)

Production occurs as a sequences of operations. Complex parts may require dozens, or even hundreds, of processing steps. The strategy of combined operations involves reducing the number of distinct production machines or workstations through which the part must be routed. This is accomplished by performing more than one operation at a given machine, thereby reducing the number of separate machines needed. Since, each machines typically involves a setup, setup time can be saved as a consequence of this strategy. Material handling effort and non-operation time are also reduced.

5.4.4 Integration of Operations

Another strategy is to link several workstations into a single integrated mechanism using automated work handling devices to transfer parts between stations. In effect, this reduces the number of separate machines though which the product must be scheduled with more than one workstation, several parts can be processed simultaneously, thereby increasing the overall output of the system.

5.4.5 Increased Flexibility

This strategy attempts to achieve maximum utilization of equipment for job shop and medium volume situations by using the same equipment for a variety of products. It involves the use of the flexible automation concepts. Prime objectives are to reduce setup time and programming time for the production machine. This normally translates into lower manufacturing lead time and lower work-in-process.

5.4.6 Automated material Handling Storage System

A great opportunity for reducing non-productive time exists in the use of automated material handling and storage systems. Typical benefits included reduced work-in-process and shorter manufacturing lead times.

5.4.7 Online Inspection

Inspection for quality of work is traditionally performed after the process. This means that any poor quality product has already been produced by the time it is inspected. Incorporating inspection into the manufacturing process permits corrections to the process as product is being made. This reduces scrap and brings the overall quality of product closer to the nominal specifications intended by the designer.

5.4.8 Process Control and Optimization

This includes a wide range of control schemes intended to operate the individual process and associated equipment more efficient. By this strategy, the individual process times can be reduced and product quality improved.

5.4.9 Plant Operations Control

Whereas the previous strategy was concerned with the control of the individual manufacturing process, this strategy is concerned with control at the plant level of computer networking within the factory.

5.4.10 Computer Integrated Manufacturing (CIM)

Taking the previous strategy, one step further, the integration of factory operations with engineering design and many of the other business functions of the film. CIM involves extensive use of computer applications, computer data bases and computer networking in the company.

5.4.11 Online Monitoring

The system here aims in implementing on online facility for monitoring our automation and the equipment or production status at any instant of demand. Detection of any error in system or product immediately show this type of monitoring is best monitoring system to reduce the non-productive, time and increase efficiency of production, also increase accuracy of the product.

Important Points

1. **Automation:**
    *   Automation can be defined as the technology used for the application of integrated mechanical, electronic and computer based systems in the operation and control of production systems.

    *   Need of automation:

        (1) To increase productivity.

        (2) To reduce cost of production.

        (3) To improve product quality.

        (4) To mitigate the effect of labour shortages.

        (5) To reduce production time.

        (6) To avoid high cost of not automating.

        (7) To have better control over manufacturing activities.

        (8) To improve worker safety.

        (9) To reduce or eliminate routine manual and clerical tasks.

2. **High and Low Cost Automation:**

   (A) High Cost Automation:

   - High cost automation involves automation of manufacturing facilities and manufacturing support systems. It demands replacement of conventional machines by CNC machine, material handling by conveyors and AGV's

   (B) Low Cost Automation:

   - Low cost automation is a technology that creates some degree of automation in some existing facilities like machine tool, material handling etc. by using standard components available in market.

3. **Elements of Automation:**

   - Power source.
   - Program of instructions.
   - Control system.
     - (a) Open loop control system.
     - (b) Close loop control system.

4. **Type of Automation:**

   - Fixed automation.
   - Programmable automation.
   - Flexible automation.

5. **Strategies in Automation:**

   - Specialization of operation.
   - Multiple operations.
   - Integration of operations.
   - Increased flexibility.
   - Automated material handling.
   - Online inspection.
   - Process control and optimization.
   - Plant operations control.
   - Computer integrated manufacturing.
   - Simplification.

Practice Questions

1. Define and need of automation.
2. Short note on high and low cost automation.
3. Explain types of automation.
4. Explain element of automation.
5. Explain fixed automation
6. Explain programmable automation
7. Explain flexible automation.
8. Comparison between fixed automation, programmable automation and flexible automation.
9. Explain different strategies used in automation.

☝ ☝ ☝

Chapter 6

ROBOT TECHNOLOGY

Weightage of Marks = 12, Teaching Hours = 08

Syllabus

6.1 **Introduction of robotics:** Definition of robot and robotics, Advantages, Disadvantages.

6.2 **Basic components of robot:** Manipulator, End effectors, Actuators, Sensors, Controller, Processor and software.

6.3 **Robot joints:** Linear, Orthogonal, Rotational, Twisting and revolving.

6.4 **Degree of freedom of robot:** Vertical, radial, Rotational traverse, Wrist pitch, Wrist yaw wrist roll.

6.5 **Actuators:** Mechanical, Hydraulic, Pneumatic and electric.

6.6 **End effectors:** Gripper and types.

6.7 **Robot sensors:** Classification of sensors.

6.8 **Basic configuration of robot:** Cartesian, Cylindrical, Polar (spherical).

6.9 **Applications of robot:** Loading unloading, Material handling, Processing operations, Assembly and inspection.

About this Chapter

At the end of this chapter, students will be able to:

- Explain with sketches the function of the specified actuators used in a robot.
- Explain given types of grippers used in robot with diagram.
- Explain with sketches the function of the given sensors used in a robot.
- Justify the use of robot in the given industrial situation.

6.1 INTRODUCTION OF ROBOTICS

6.1.1 Definition of Robot and Robotics

- **Define robot:** The robot institute of America defines the **robot** as, *"a reprogrammable, multi-functional manipulator designed to move materials, parts, tools or specialized devices through variable programmed motions for performing a variety of tasks".*

- **Define robotics:** It is defined as the combination of machine tool technology and computer science and is form of industrial automation and is a technology with a future and for a future.

6.1.2 Advantages of Robots

- The robots are the important parts of the modern days industry. The robots offer the following advantages:

    1. Robots increase productivity, safety and efficient of process.

    2. Robots improve the quality and consistency of work.

    3. Robots can work in hazardous environments without the need for life support.

    4. Robots need no environmental comfort, such as : lighting, air conditioning, ventilation and noise protection.

5.   Robots work continuously without experiencing fatigue or boredom.

6.   Robots have repeatable precision at all times.

7.   Robots can operate with high degree of accuracy.

8.   Robots can have capabilities beyond that of humans.

9.   Robots can process multiple tasks simultaneously.

6.1.3 Disadvantages of Robots

- The robots have certain inherent disadvantages. The disadvantages of robots area as follows:

(1)   Robots replace human workers creating economic problems, such a lost salaries; and social problems, such as dissatisfaction and resentment among workers.

(2)   Robots have limited capabilities in :
   - degrees of freedom,
   - dexterity,
   - sensors,
   - vision system and
   - real-time response.

(3)   Robots are costly due to:
   - high initial investment,
   - installation costs,
   - peripherals cost,
   - training cost, and
   - programming cost.

(4)   Robots lack capability to respond in unpredictable emergencies.

6.2 BASIC COMPONENTS OF ROBOTS

A typical robot, shown in Fig. 6.1, consists of following components:

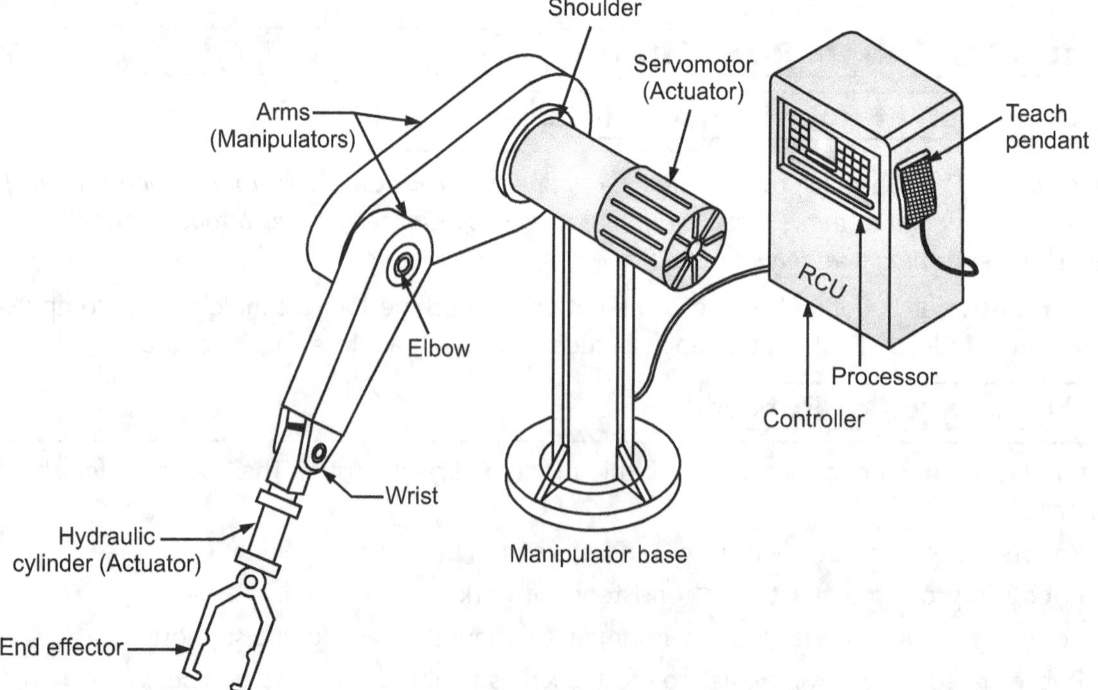

Fig. 6.1: Typical Robot

6.2.1 Manipulator

- Manipulator is the combination of mechanical linkages, connected by joints to from an open-loop kinematic chain.

- The manipulator is capable of movement in various directions. The joints of the manipulator produce the motion which is either rotary or linear.

- The manipulator gets the task performed through the end effector, which is connected to the manipulator.

6.2.2 End Effector

- The end effector is the part that is connected to the last joint (hand) of a manipulator, which generally performs the required tasks or handles the objects.

- The hand of a robot has provisons for connecting the end effector that is specially designed for a purpose.

- The end effector is either controlled by the robot's controller communicates with the end effector's controlling device such as PLC.

6.2.3 Actuators

- The actuators are the drives used to actuate the joints of the manipulator. They produce relative rotary or linear motion between the two links of joint. In short, they are the 'muscles' of the manipulator.

- The common types of actuators are: servomotors, stepper motors, pneumatic cylinders and hydraulic cylinders.

- The actuators are controlled by controller.

6.2.4 Sensors

- The sensors are used to collect the information about the status of the manipulator and the end effector. This can be done continuously or at the end of a desired motion.

- The sensors are used to collect the information about the status of the manipulator and the end effector. This can be done continuously or at the end of a desired motion.

- The sensors collect the information like: instantaneous position, velocity and acceleration, or various links and joints of the manipulator.

- The information is sent to the controller using the information, the controller determines the configuration of the robot and controls the movement of the manipulator.

- The information sent by the sensors can be either analog digital or combination of two.

- The sensors used in robots can be dividend into two classes:

  (1) Non-visual sensors.

  (2) Visual sensors.

  (1) Non-visual sensors: The non-visual sensors include: limit switches, position sensors, velocity sensors or force and tactile sensors.

  (2) Visual sensors: The visual sensors include: TV cameras, vision system, charge-coupled device (CCD), or charge injection device (CID).

6.2.5 Controller

- The controller receives the instructions from the processor of a computer and controls the motion of the actuators. It takes feedback from the sensors.
- The controller performs the following three functions:
    (i) It stores the position and sequences date of the manipulator.
    (ii) It initiates and terminates the motion of the individual components (links) of the manipulator in a desired sequence and at the specified point.

6.2.6 Processor

- The processor is the brain of the robot, which calculates the motion of the joints so as to achieve the desired action of the robot.
- It sends signals to the controller and receives the feedback from the controller.
- The processor is a computer which is dedicated to a single purpose.

6.2.7 Software

- There are generally three groups of software that are used in robot:
    (i) Operating system: For opening the computer.
    (ii) Robotic software: For operation of the robot.
    (iii) Application programmes: For operation of peripheral devices.

6.3 ROBOT JOINTS

6.3.1 Linear Joints

- The linear joint provides the translational sliding motion between the input and output link.
- The axes of both are links are parallel to one another.
- Such a joint is called as 'type L' joint.
- The linear joint is shown in Fig. 6.2.

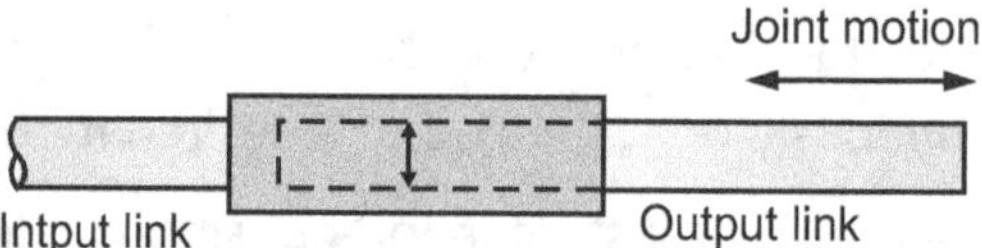

Fig. 6.2: Linear Joint

6.3.2 Orthogonal Joint

- The orthogonal joint provides the translational sliding motion between the input and output link.
- The axis of output link is perpendicular to that of the input link.
- The orthogonal joint is shown in Fig. 6.3.

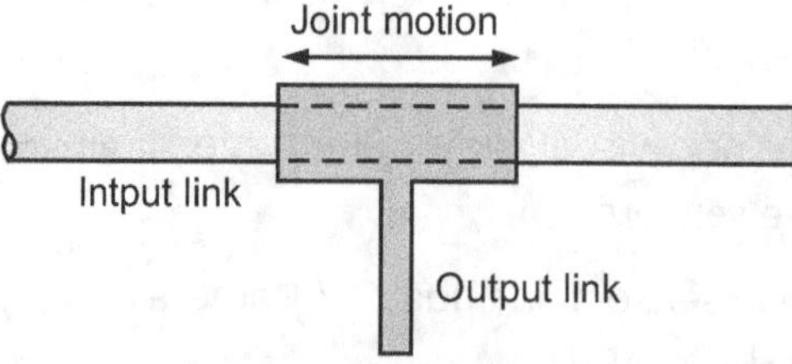

Fig. 6.3: Orthogonal Joint

6.3.3 Rotational Joint

- The rotational joint provides the relative rotational motion between the input and output link.
- The axis of rotation is perpendicular to the axes of input and output link.
- Such a joint is called as 'type R' joint.
- The rotational joint is shown in Fig. 6.4.

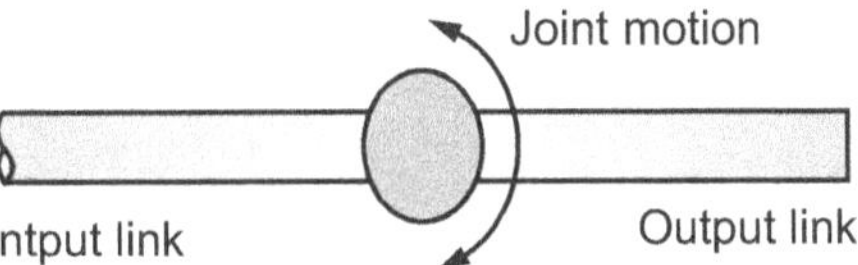

Fig. 6.4: Rotational Joint

6.3.4 Twisting Joint

- The twisting joint provides the relative rotational motion between the input and output link.
- The axis of rotation is parallel to the axes of input and output link.
- Such a joint is called 'type T' joint.
- The twisting joint is shown in Fig. 6.5.

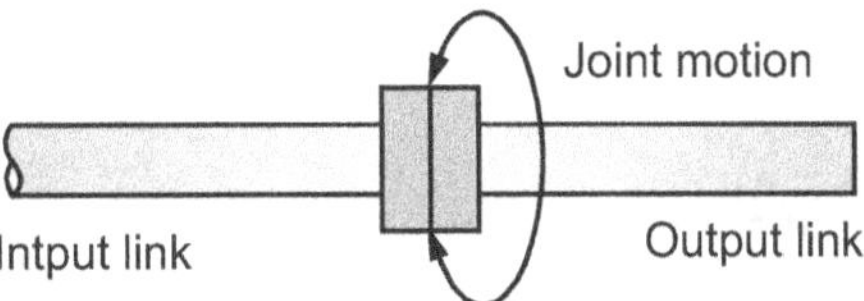

Fig. 6.5: Twisting Joint

6.3.5 Revolving Joint

- The revolving joint provides the relative rotational motion between the input and output link.
- The axis of the input link is parallel to the axis of rotation of the joint.
- The axis of the output link is perpendicular to the axis of rotation of the joint.
- Such a joint is called as 'type V' point.
- The revolving joint is shown in Fig. 6.6.

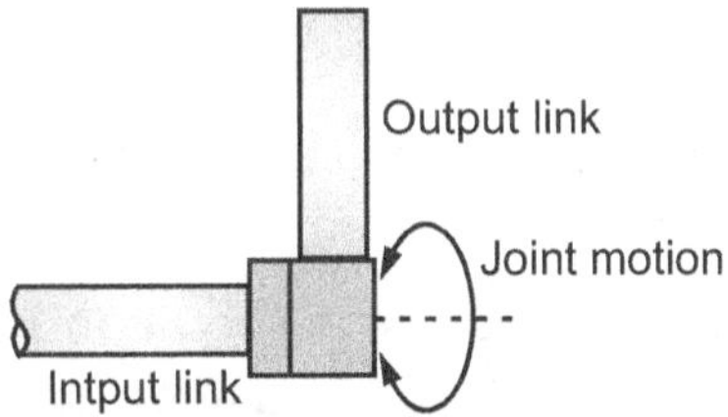

Fig. 6.6: Revolving Joint

6.4 DEGREE OF FREEDOM OF ROBOT

- Every joint has one Degree Of Freedom (D.O.F.). Hence, total DOFs of robot is equal to the number of joints.
- Many robots have six DOF's - three rotational for orientation in space and three translational for positioning.
- However, it is possible to have as few as two and as many as eight DOFs.

- A robot having six degrees of freedom is shown in Fig. 6.7. The six degrees of freedom (DOFs) are as follows:

    (I) Degrees of Freedom of Arm:

    1. Vertical Traverse

    2. Radial Traverse

    3. Rotational Traverse

    (II) Degrees of Freedom of Wrist:

    (4) Wrist Pitch

    (5) Wrist Yaw

    (6) Wrist Roll.

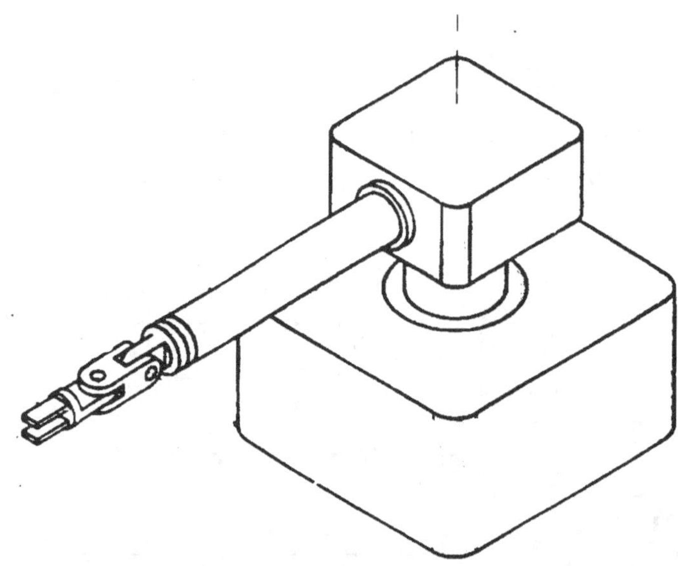

Fig. 6.7: Robot with Six Degrees of Freedom

(I) Degrees of Freedom of Arm:

The arm typically has three degrees of freedom:

(1) **Vertical Traverse:** Vertical traverse is the upward or downward movement of the arm. This movement allows the robot to cover the height during its operation.

(2) **Radial Traverse:** Radial traverse is the in and out movement of the arm along its axis. This movement allows the area during its operation.

(3) **Rotational Traverse:** Rotational traverse is the rotation of the arm about the vertical axis. this movement allows the robot to occupy the desired angular position about the vertical axis.

(II) Degrees of Freedom of Wrist:

The wrist has three degrees of freedom:

(4) **Wrist Pitch:** Wrist pitch is the up and down rotation or pitching of the wrist.

(5) **Wrist Yaw:** Wrist yaw is the rotation of the wrist in horizontal plane about the vertical axis of the wrist.

(6) **Wrist Roll:** Wrist roll is the rotation or rolling motion of the wrist about its longitudinal axis.

6.5 ACTUATORS (DRIVES) FOR ROBOTS

Actuators are the devices which provides the actual motive force for the manipulator joints of the robots.

Based on the types of actuating elements, the actuators are classified into four types:

(1)　Mechanical actuators.

(2)　Hydraulic actuators.

(3)　Pneumatic actuators.

(4)　Electric actuators.

6.5.1　Mechanical Actuators

- The mechanical actuators use the mechanical elements like: rack and pinion, gears, power screws, belts etc. For providing the motive force/torque for the manipulator joints of the robots.

- Based on the type of movement or motion, the mechanical actuators are further classified into two types:

　(i)　Linear Mechanical Actuators.

　(ii)　Rotary Mechanical Actuators.

　(i)　Linear Mechanical Actuators: The following linear mechanical actuators are used to actuate the linear manipulator joints of the robots:

　　(a)　Rack and pinion.

　　(b)　Recirculating ball screws.

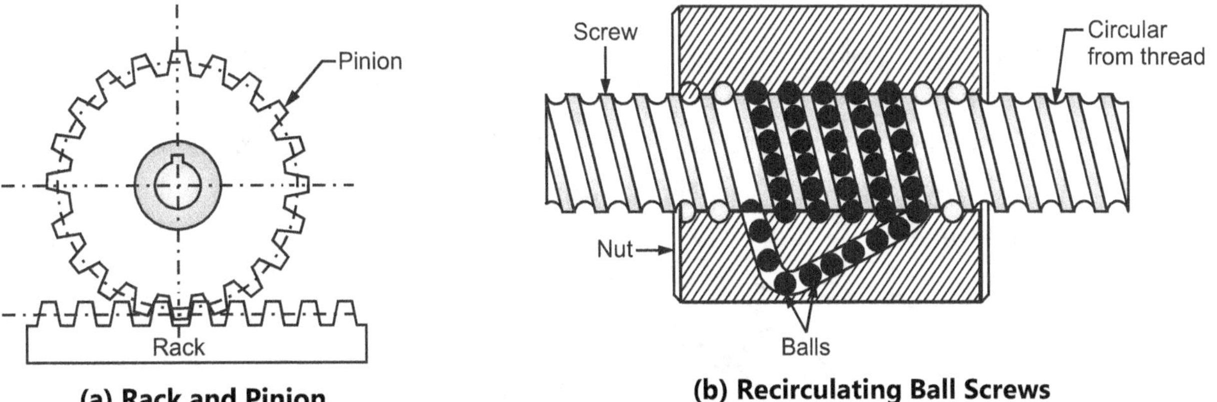

(a) Rack and Pinion　　　　　**(b) Recirculating Ball Screws**

Fig. 6.8

　(ii)　Rotary Mechanical Actuators:

The following rotary mechanical actuators are used to actuate the rotary joints of robots:

(a)　Timing belts.

(b)　Gear pairs.

(c)　Harmonic drives.

(a)   Timing belts:

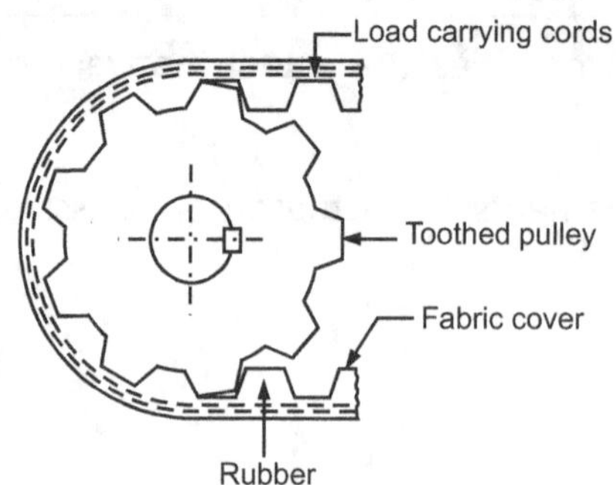

Fig. 6.9 (a): Timing Belts

(b)   Gear pairs:

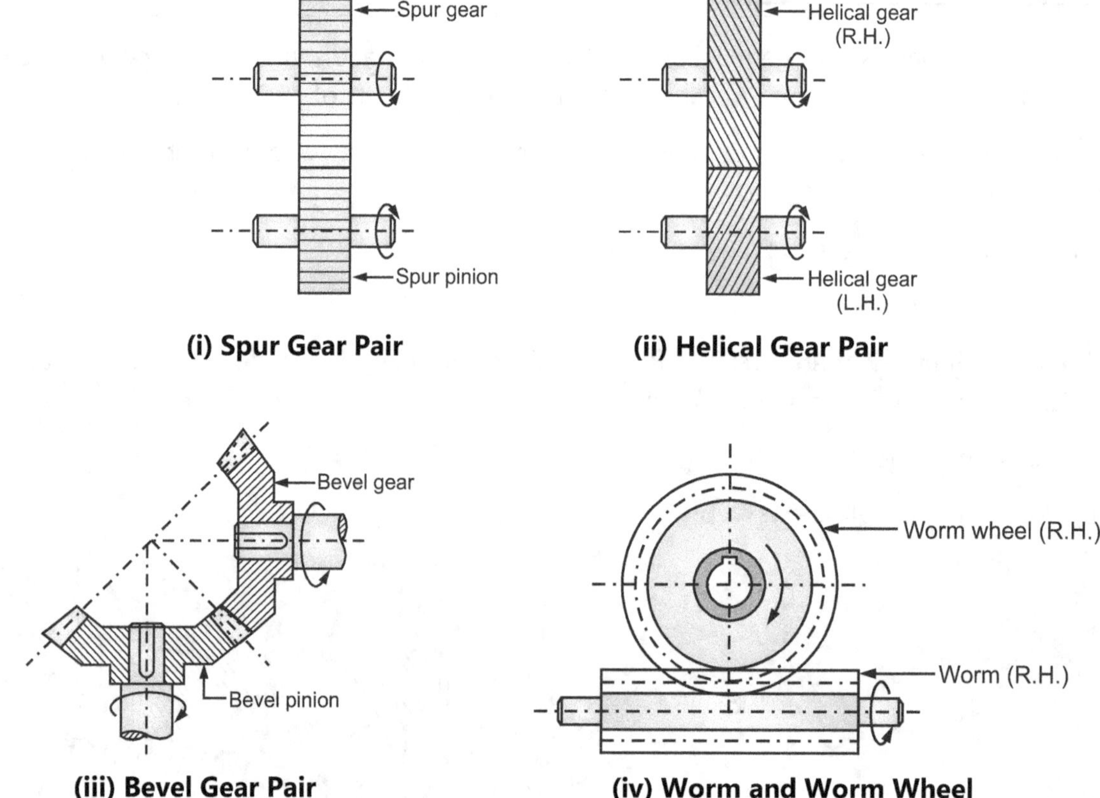

(i) Spur Gear Pair   **(ii) Helical Gear Pair**

(iii) Bevel Gear Pair   **(iv) Worm and Worm Wheel**

Fig. 6.9 (b): Gear Pairs

(c)   Harmonic drives:

- Harmonic drive is a compact arrangement which gives high reduction ratio (15:0 to 1:100)

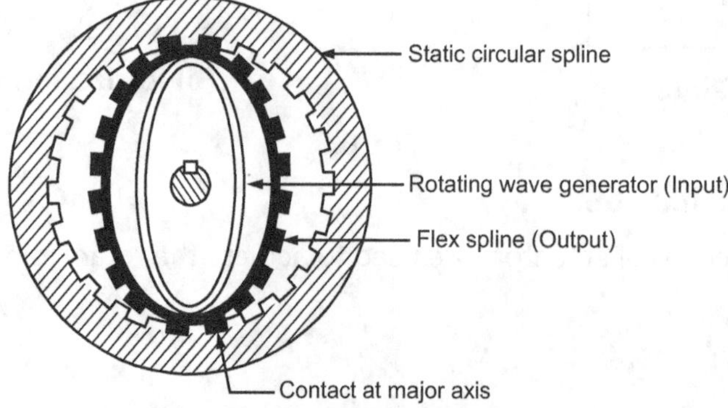

Fig. 6.10: Harmonic Drive

- It eliminates backlash and tooth wear which is present in conventional reduction gear box. Because of its smooth, backlash free and efficient action, it is suitable for robot drives.

- The hydraulic actuators uses the pressurized fluid for providing motive force/torque for the manipulator joint of the robots.

- The hydraulic actuators are further classified into two types:

(i)   Linear hydraulic actuators.

(ii)   Rotary hydraulic actuators.

(i)   Linear Hydraulic Actuators: The linear hydraulic actuators are the single acting or double acting hydraulic cylinders used to actuate the linear joints by means of piston rod movement. The high pressure hydraulic oil at 70 bar at 210 bar pressure operates the piston and hence the piston rod.

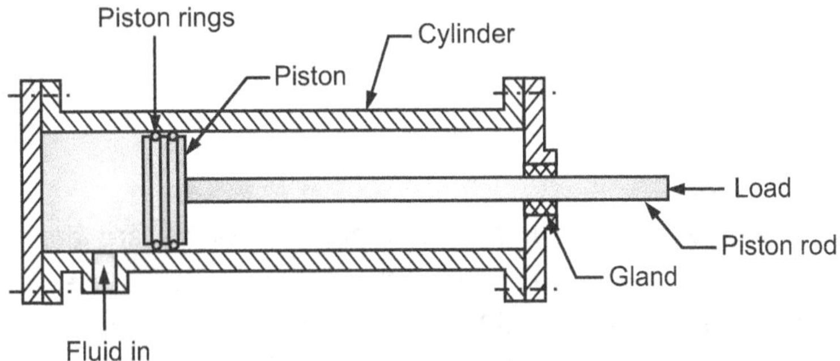

(a) Single Acting Hydraulic Cylinder

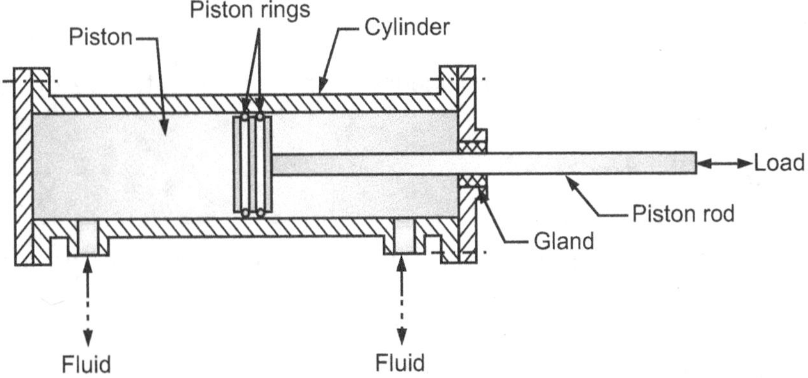

(b) Double Acting Hydraulic Cylinder

Fig. 6.10: Linear Hydraulic Actuators

(ii)   Rotary Hydraulic Actuators: There are three types of rotary hydraulic actuators.

(a) Gear motors,

(b) Vane motors,

(c) Piston motors.

The high pressure hydraulic oil (70 bar to 210 bar) runs the hydraulic motors (gear motors, vane motors, or piston motors). The torque developed by hydraulic motor actuates the rotary joints.

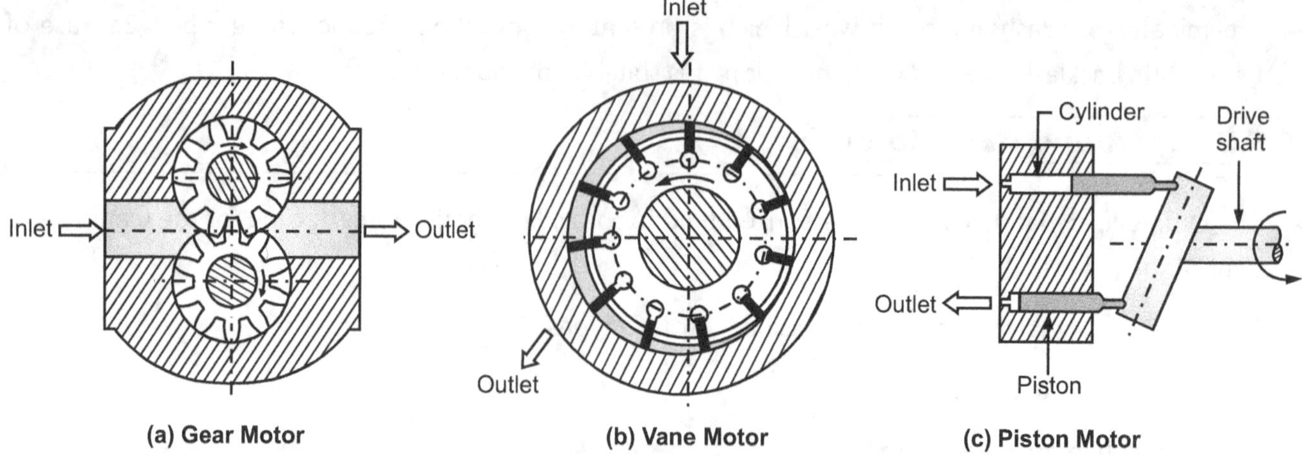

Fig. 6.11: Rotary Hydraulic Actuators

6.5.3 Pneumatic Actuators

- The pneumatic actuators use the compressed air for providing motive force/torque for the manipulator joints of the robots.

- The pneumatic actuators are further classified into two types:

    (i)　Linear Pneumatic Actuators.

    (ii)　Rotary Pneumatic Actuators.

    (i)　Linear Pneumatic Actuators: The linear pneumatic actuators are the single acting or double acting pneumatic cylinders used to actuate the linear joints of the robots. The cylinders are operated by compressed air at about 35 to 70 bar pressure.

    (ii)　Rotary Pneumatic Actuators: There are two types of rotary pneumatic actuators:

    (a) Vane motors,

    (b) Piston motors.

The high pressure compressed air runs the pneumatic motors. The torque developed by pneumatic motor actuates the rotary joints.

6.5.4 Electric Actuators

- The different types of electric motors are used as rotary actuators in robots. They are:

    (i)　D.C. motors.

    (ii)　Reversible A.C. motors.

    (iii)　Brushless D.C. motors.

    (iv)　D.C. servomotors.

    (v)　A.C. servomotors.

    (vi)　Stepper motors.

- A stepper motor is a unique type of rotary actuator which provides rotation in the form of discrete steps of fixed angular displacement.

6.5.5 Comparison of Electric, Hydraulic and Pneumatic Actuators

- Comparison of electric hydraulic and pneumatic actuators is given in Table 6.1.

Table 6.1: Comparison of electric, hydraulic and pneumatic actuators

Sr. No.	Comparison parameter	Electric actuators	Hydraulic actuators	Pneumatic actuators
1.	Payload capacity	Electric actuators are suitable for robots with moderate payload capacity.	Hydraulic actuators are suitable for robots with high payload capacity.	Pneumatic actuators are suitable for robots with low payload capacity.
2.	Power to weight ratio	Electric actuators have high power to weight ratio.	Hydraulic actuators have moderate power to weight ratio.	Pneumatic actuators have low power to weight ratio.
3.	Accuracy and precision	Electric actuators are highly accurate and precise.	Hydraulic actuators have moderate accuracy and precision.	Pneumatic actuators have relatively low accuracy and precision due to compressibility of air.
4.	Compatibility with electronic controller	Electric actuators are highly compatible with electronic controller.	Compatibility of hydraulic actuators with electronic controller is not as good as that of electric actuators.	Compatibility of pneumatic actuators with electronic controller is not as good as that of electric actuators.
5.	Reliability and maintenance	Electric actuators are highly reliable and require low maintenance.	Due to leakage problem, hydraulic actuators have low reliability and require high maintenance.	Pneumatic actuators have moderate reliability and require moderate maintenance.
6.	Cleanliness and quietness of operation	The operation of electric actuators are clean and quiet.	Due to leakage, the operations of hydraulic actuators are not clean. In addition, the operation are noisy.	The operation of pneumatic actuators are clean but noisy.
7.	Requirement of transmission elements	Electric actuators require transmission elements like: gears, rack and pinion, belts etc. This increases the cost of the system.	Hydraulic actuators do not require extra transmission elements.	Pneumatic actuators also do not require extra transmission elements.
8.	Comapctness	Electric actuators are highly compact.	Hyraulic actuators need auxiliary equipment like: motor, pump, reservoir, hoses etc. Hence, system is bulky.	Pneumatic actuators need auxiliary equipment like: motor, compressor, hoses, system is bulky. But it is not as bulky as hydraulic actuators system.
9.	Cost of system	Low cost.	High cost.	Moderate cost.
10.	Operational speed range	Electric actuators can work in narrow range of speeds.	Hydraulic actuators can work in moderate range of speed.	Pneumatic actuators can work in wide range of speed.
11.	Ability to withstand shock	Electric actuator cannot with stand heavy shocks.	Hydraulic actuators can withstand heavy shocks.	Due to compressibility of air, pneumatic actuators are not suitable for shock loads.

... (Contd.)

Sr. No.	Comparison parameter	Electric actuators	Hydraulic actuators	Pneumatic actuators
12.	Working environment	Due to fire risk, electric actuators are not suitable in explosive as well as wet environment.	Hydraulic actuators are suitable in explosive environment but not suitable in wet environment.	Pneumatic actuators are suitable in both explosive as well as wet environment.
13.	Breaking requirement	Electric actuators require braking device, when not powered; otherwise the arm will fall.	Hydraulic actuators do not require braking device, when not powered.	Pneumatic actuators require braking device powered.

6.6 END EFFECTORS OF ROBOT

- **End effector** is a device that is attached to the wrist of the robot arm so as to enable the robot to perform a specific task. It is, sometimes, referred as the 'hand' of the robot.

- Due to diverse applications of robots, end usually customized for a particular application.

- Sometimes, the robot handles the part which changes its shape during a process. In such cases, the end effectors are designed as multifunctional.

6.6.1 Types of End Effects

The end effectors are broadly classified into two types:

(1) Grippers,

(2) Tools.

(1) Grippers:

- Grippers are the end efforts used for holding the parts or objects.

- The part handling applications include: machine loading and uploading, picking and placing of parts on conveyor, arranging parts onto a pallet etc.

(2) Tools:

- In many applications, robot is required to operate tools rather than handling the parts. In such cases, tools are used as the end effectors.

- The examples of tools used as end effectors are: spot welding tool, arc-welding torch, spray paining nozzle, wrench, machining tools etc.

6.6.2 Grippers

- As discussed earlier, grippers are the end effectors used for holding the parts or objects.

- Some of the applications of grippers are machine loading and unloading, picking and placing of parts on conveyor, material handling, bottle handling, arranging parts onto pallets etc.

6.6.3 Considerations in Selection of Grippers

- The various factors to be considered while selecting the gripper for a given application are as follows:

  (1) Type of power source: The type of power source available (i.e. electric hydraulic or pneumatic) is an important parameter in the design and selection of gripper for robot.

(2) Method of actuation: The different methods used for graping the objectives or parts are:

- Mechanical grasping.

- Magnetic grasping.

- Vacuum grasping.

- Adhesive grasping.

- Expanded bladder type grasping.

- Other methods (hooks, scoops etc.)

(3) Weight of part to be handled: The weight of the part to be handled influences the size of the gripper and the required actuating force.

(4) Configuration of part to be handled: The size, shape, tolerances, surface finish and delicacy of the part to be handled must be considered while designing the gripper.

(5) Change of configuration of part during process: The change in size, shape, delicacy, surface finish and hardness of the part between loading and unloading time must be considered while designing the gripper.

(6) Material of part to be handled: The type of the material of the part to be handled plays an important role in designing and selecting the gripper. It is important to know whether the material is rigid or flexible, ductile or brittle, hard or soft.

(7) Part fixture: The type of part fixture decides the required accuracy and precision of the gripper addition, it also decides the method of gripping and gripping force required.

(8) Cycle time and number of actuations per day: The cycle time and number of actuations of gripper per day are important consideration in deciding the type of gripping.

(9) Characteristics of robot: The different characteristics of robot to be considered in gripper design are: size and shape of work envelope payload, accuracy, precision, reach and construction.

(10) Operating environment: The various operating environment factors like: temperature, humidity, moisture, chemicals, dirt etc. are considered while designing the gripper.

(11) Multifunctionality requirement:

- Sometimes, the grippers are required to perform the multiple tasks.

- However, this may increase the complexity, increase the cost and reduce the reliability.

(12) Location of sensor:

- Sometimes, the sensor is located on gripper itself.

- In such cases, the design of gripper should be such that it gives due protection to the sensor.

(13) Workplace (floor) layout: The layout of workplace and space available for operations should be considered while designing the gripper.

(14) Simplicity and serviceability: The gripper should be simple in design and easy for servicing. It should use minimum possible parts.

(15) Cost of gripper: The gripper should be cost effective.

6.6.4 Types of Grippers

Fig. 6.12 shows the different types of grippers.

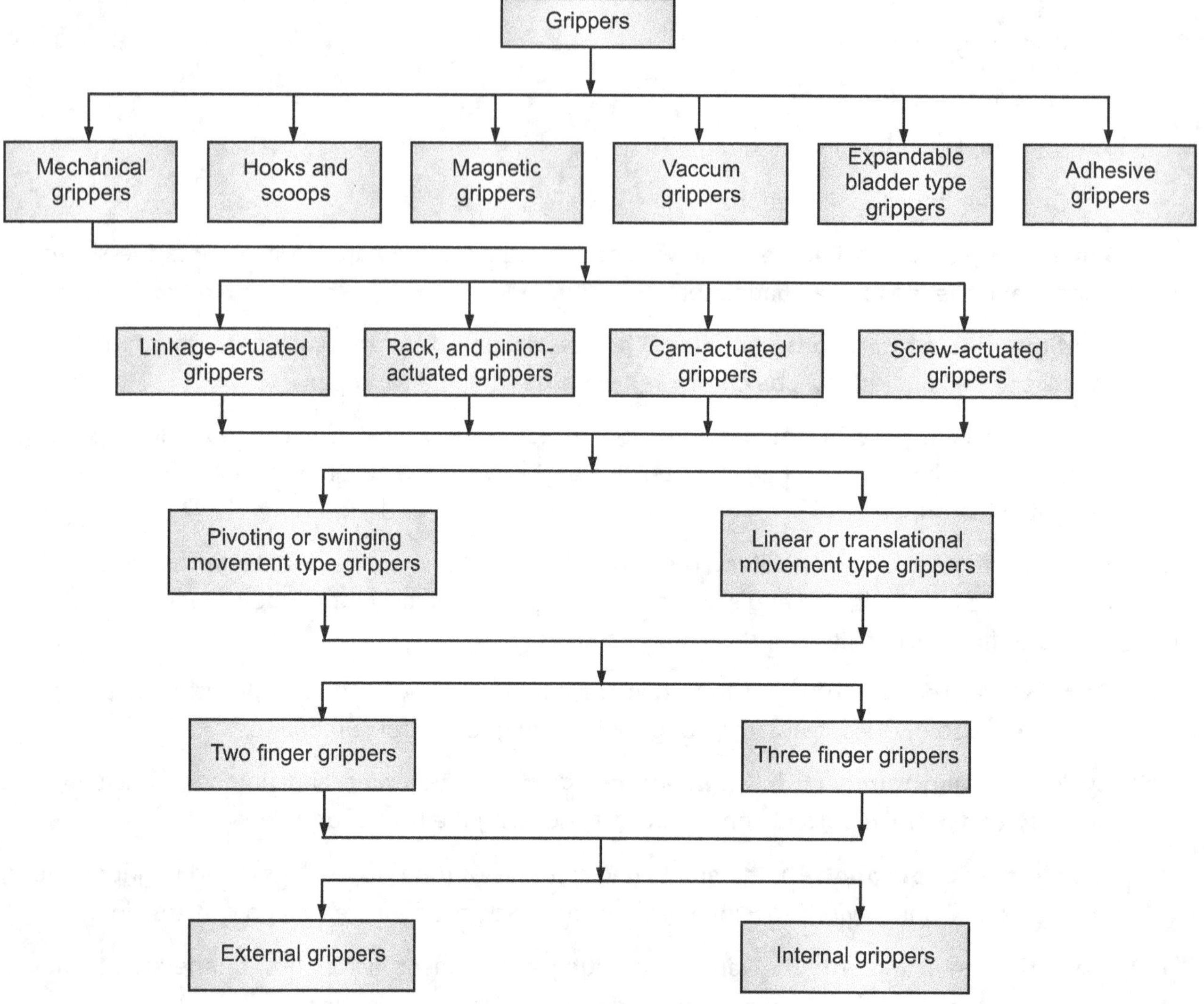

Fig. 6.12: Types of Grippers

Based on the method of grasping the parts or objects, the grippers are classified into following six types:

1. Mechanical grippers.

2. Hooks and scoops.

3. Magnetic grippers.

4. Vacuum grippers.

5. Expandable bladder type grippers.

6. Adhesive grippers.

6.6.4.1 Mechanical Grippers

- The mechanical grippers use the mechanical fingers actuated by a mechanism to grasp an object. the fingers are either attached to the mechanism or are an integral part of the mechanism.

- The attached fingers, shown in Fig. 6.13 are more convenient because they are replaceable and

- In order to accommodate different part models, different sets of fingers can be used with the same gripper mechanism.
- The function of the gripper mechanism is to translate some form of power input the grasping action of the fingers. The power input can be mechanical, electrical, hydraulic or pneumatic.

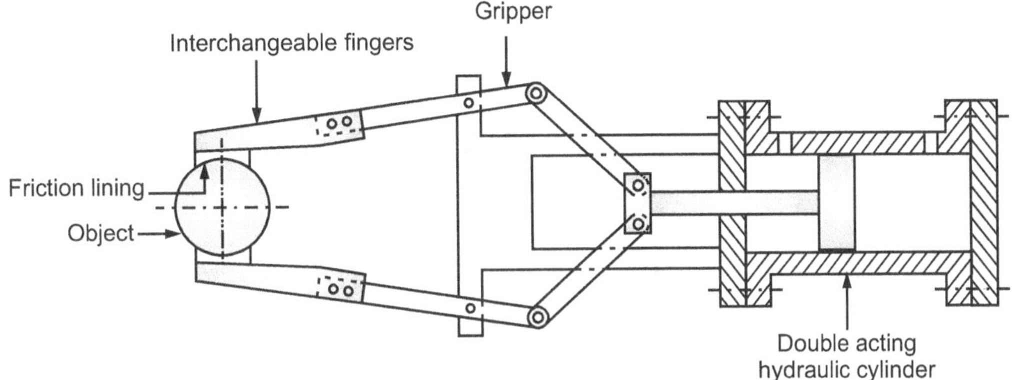

Fig. 6.13: Mechanical Gripper with Interchangeable Fingers

6.6.4.2 Hooks and Scoops

- Hooks can be used as grippers to load and unload part hanging from the overhead conveyors.
- The part to be handle by a hook must have some sort of handle, eye bolt or ring to enable the hook to hold it.
- Scoops are used as end efforts to handles the materials in liquid or powder form.
- The limitation of scoop is, it is difficult to control the amount of material being handled by the scoop.
- In addition, spilling of the material during handling is another problem.

6.6.4.3 Magnetic Grippers

- Magnetic grippers can be used for handling ferrous materials.
- **Advantages of magnetic grippers:**

  (a) Variations in part size can be tolerated;

  (b) They can handle metal parts with holes;

  (c) They require only one surface for gripping; and

  (d) Picking time is very fast.

- **Limitations of magnetic grippers:**

  (a) Residual magnetism remaining in the workpiece may cause problems.

  (b) It is difficult to pickup only one sheet from the stack. The magnetic attraction tends to penetrate beyond the top sheet in the stack, resulting in the possibility that more than a single sheet will be lifted by the magnet.

- **Types of magnetic grippers:**

  The magnetic grippers are further divided into two types:

  (i) Electromagnetic grippers: The electromagnetic grippers require D.C. power source and controller unit. The electromagnetic grippers are easy to control. In order to release the part, the controller unit reverses the polarity at a reduced power level before switching off the electromagnet.

(ii) Permanent magnetic grippers: Permanent magnetic grippers do not require external power source, and hence they can be used in hazardous and explosive environment.

However, in case of permanent magnetic grippers, in order to release the part, some means of separating the part from the magnet must be provided. [Fig. 6.14]

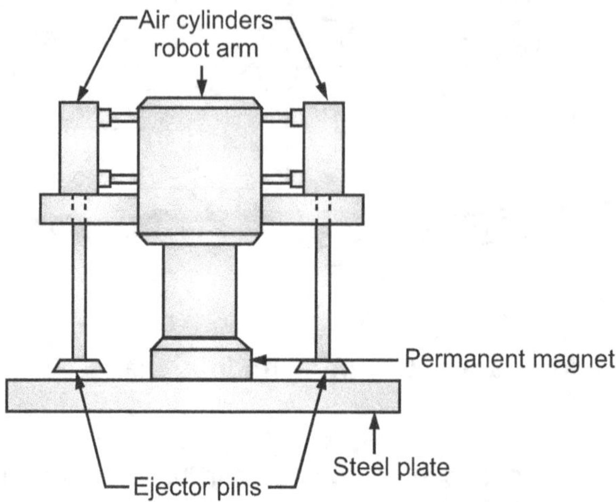

Fig. 6.14: Permanent Magnetic Gripper

6.6.4.4 Vacuum Grippers

- Large flat and smooth objects are difficult to grasp. Vacuum cups can be used as gripper devices for picking up: flat and smooth metal plates, glass pans, large light weight boxes etc.

- The vacuum cups are operated by vacuum pump. The capacity of vacuum gripper mainly depends upon the negative pressure created by vacuum pump.

- Fig. 6.15 shows a typical vacuum gripper.

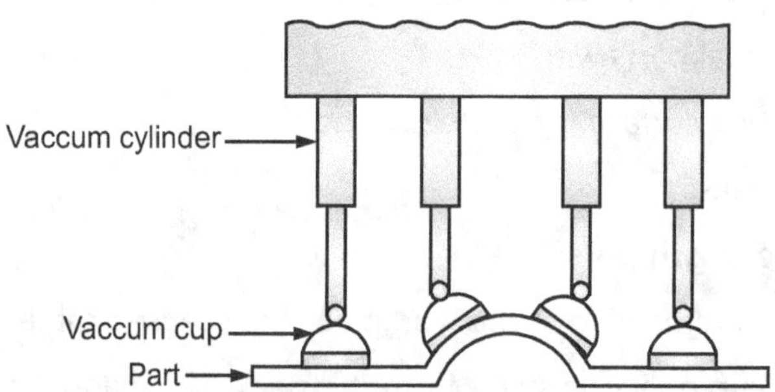

Fig. 6.15: Vacuum Gripper

6.6.4.5 Expandable Bladder Type Grippers

- As mechanical grippers apply concentrated force, they are not suitable for grasping the fragile objects.

- The expandable bladders fabricated out of rubber or other elastic material can be used for gripping fragile objects from internal surface.

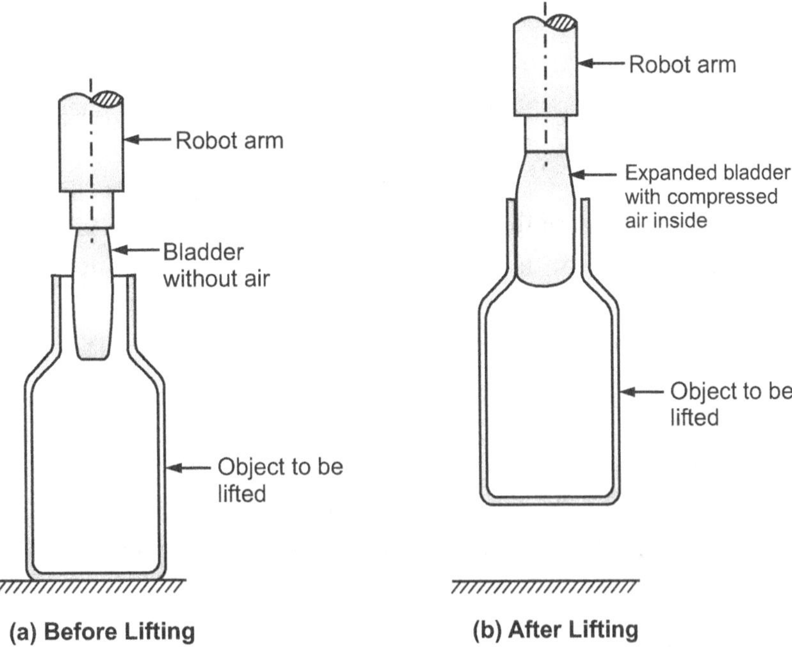

Fig. 6.16: Expandable Bladder Type Grippers

6.6.4.6 Adhesive Grippers

- Adhesive substance can be used for grasping action in a gripper to handle fabrics and other light weight materials.

- In adhesive grippers, the adhesive substance losses its tackiness due to repeated usage. This reduces the reliability of the gripper.

- In order to overcome this difficult, the adhesive material is continuously fed to the gripper in the form of ribbon by feeding mechanism.

6.7 ROBOT SENSORS

- Transducer is a device that converts one type of physical variable (e.g. force, pressure, temperature, velocity, flow rate etc.) into another form, commonly electrical voltage.

- Sensor is a transducer used to make a measurement of physical variable.

- Any sensor requires calibration in order to be useful as a measuring device. The calibration is the procedure by which the relationship is established between the measured variable and the converted output signal.

- The sensors are used to collect the information about the status of the manipulator and end effector with respect to object position. This can be done continuously or at the end of a desired motion.

- The sensors collect information like instantaneous position. Velocity and acceleration of various links and joints of the manipulator.

- This information is sent to the controller. Using this information, the controller determines the configuration of the robot at the given instant and controls the movement of the manipulator.

- The information sent by the sensors can be either analog, or combination of two.

6.7.1 Desirable Features of Sensors

- In order to be useful as a meaning devices in robots, the sensors must possess the following features:

    1. **Accuracy:** The accuracy of the measurement should be as high as possible. The output of sensor should truly reflect the input quantity measured by the sensor. The average error between the actual value and the value measured by the sensor should tend to zero.

2. **Precision:** The precision of the measurement should be a high as possible.

3. **Operating range:** The sensor should possess a wide operating range.

4. **Speed of response:** The sensor should be capable of responding to changes in the sensed variable minimum time. Ideally, the response should be instantaneous.

5. **Sensitivity:** The sensitivity of the sensor is the change in output exhibited by the sensor for a unit change in input. The sensitivity of the sensor should be as high as possible.

6. **Linearity:** The relation between the output of sensor and the input quantity measured by the sensor should be linear.

7. **Interfacing:** The sensor should be easy to interface with other devices such as: microprocessors and controllers.

8. **Calibration:** The sensor should be easy to calibrate.

9. **Reliability:** The sensor should possess a high reliability.

10. **Cost and ease of operation:** The cost of purchase installation and operation of sensor should be as low as possible. The installation and operation of sensors should be simple and easy.

11. **Size and weight:** The size and weight of the sensor should be as small as possible.

6.7.2 Types of Sensors

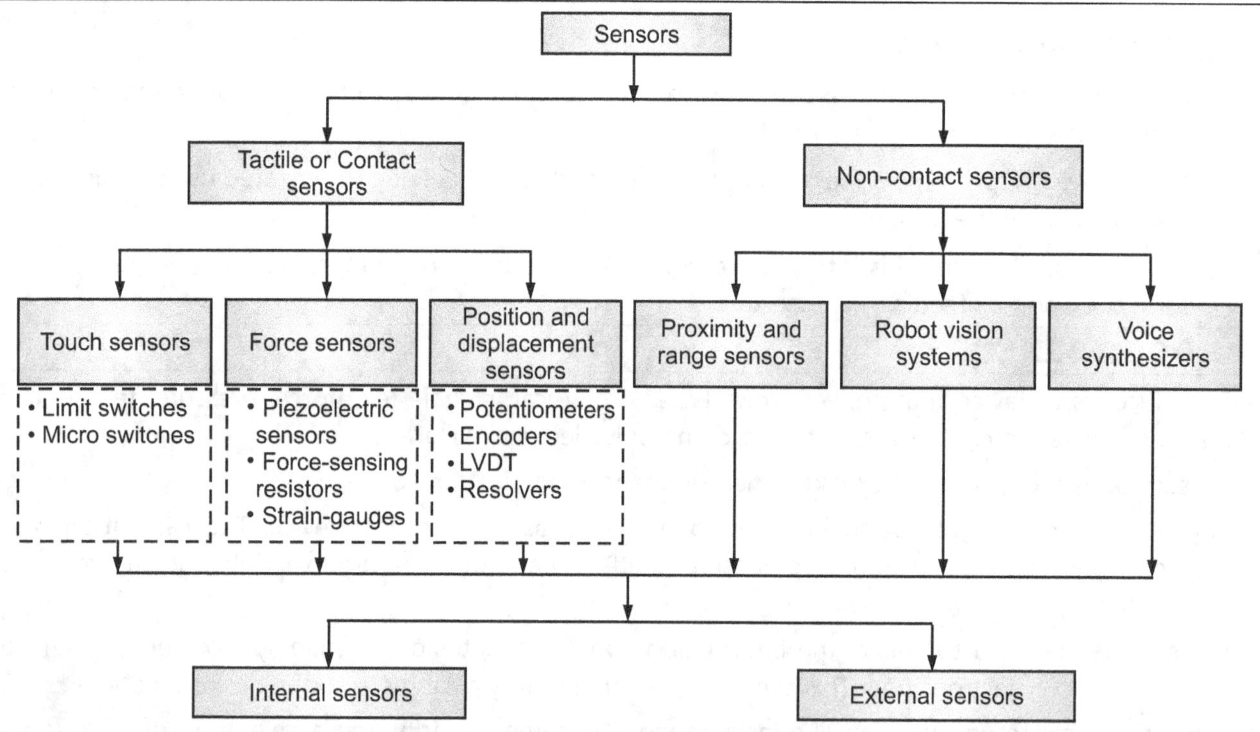

Fig. 6.17: Types of Sensors

6.7.2.1 Classification Based on Contact

Based on the contact between the sensor and the object, the sensors are broadly classified into following two types:

(1) Tactile or contact sensors:

- Tactile sensors measure the parameters by making the physical contact with the object.
- Tactile or contact sensors are further sub-dividend into three categories.

(i) Touch sensors:

- Touch sensors, also called as binary sensors, are used to indicate whether the contact has been made with the object or not, without regard to the contacting force.

- These sensors indicates the respond to the presence or absence of an object. They provide binary output signals.
- Limit switches and micro-switches are the examples of touch sensors.

(ii) Force sensors:

- Force sensors are analog type sensors in which the output signal is proportional to a local force.
- Force sensors indicates the magnitude of the contact force between the object and the sensor.
- The example of the force sensors are: piezoelectric sensors, force - sensing resistor, strain gauges etc.

(iii) Position and displacement sensors:

- The position and displacement sensors are used to measure the displacement, both rotary and linear.
- The examples of the displacement sensors are: potentiometers, encoders, liner variable differential transformers (LVDT), resolves etc.

(2) Non-contact Sensors:

- Non-contact sensors measure the parameters without contacting the object.
- Non-contact sensors are further sub-dividend into three categories.

(i) Proximity and range sensors:

- Proximity sensors give an indication, when object is close to the sensor.
- The distance required to activate the sensor can be any where between several millimeters and several meters and is dependent of the type of sensor.
- Range sensors are used to measure the distance between the object and the sensor.
- The various techniques used in proximity and range sensors are: optical devices, acoustics, eddy currents, magnetic fields etc.

(ii) Robot (or machine) vision systems:

- Robot (or machine) vision system is concerned with the sensing the three dimensional vision data and its interpretation by a computer.
- A typical robot vision system consist of: a camera, a digitizing hardware, a digital computer and a software.
- **Functions of robot vision systems:** The operation of robot vision system consist of three functions:
  - **(a) Sensing and digitizing image data:** In this process the visual image taken by camera is typically digitized and stored in the computer memory.
  - **(b) Image processing and analysis:** The digitized image is subjected to image processing and analysis.
  - **(c) Application:** The applications of the robot vision system include: inspection, part identification, location and orientation.

(iii) Voice synthesizers:

- Voice synthesis includes voice sensing and voice programming.
- Voice programming is used for oral communication of instruction to the robot. Voice sensing relies on the techniques of speech recognition to analyze spoken words uttered by a human and compare those words with a set of stored word patterns. When the spoken word matches the stored word, robot performs the particular action which corresponds to the word.

6.7.2.2 Classification Based on Reference Position

Based on the reference position with respect to which the parameters are measured by sensors, the sensors are classified into two types:

(1) Internal sensors: Internal sensors are used for measurement of parameter with respect to some reference position on robot itself.

(2) External sensors: External sensors are used for measurement of parameters with respect to some reference position outside the robot structure.

6.8 BASIC CONFIGURATION OF ROBOT

- Work envelope or work volume of a robot can be defined as the space within which the end effector of the robot can operate or reach.

- Based on the co-ordinate system of motion of the manipulator and end effector, there are four basic configurations of robots.

6.8.1 Cartesian Configuration Robots

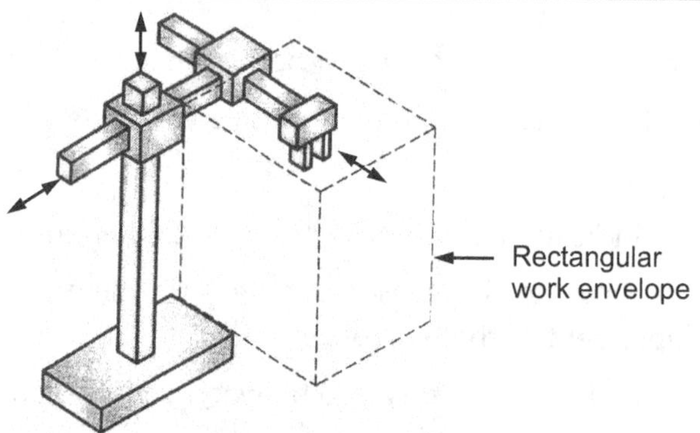

Fig. 6.18: Cartesian Configuration Robot

- Cartesian configuration robot, shown in Fig. 6.18, provides three linear motions along three mutually perpendicular axes: Y, X and Z. However, there is no rotary motion.

- This configuration provides rectangular work envelope.

- Cartesian configuration robots are used for assembly palletizing and machine tool loading.

6.8.2 Cylindrical Configuration Robots

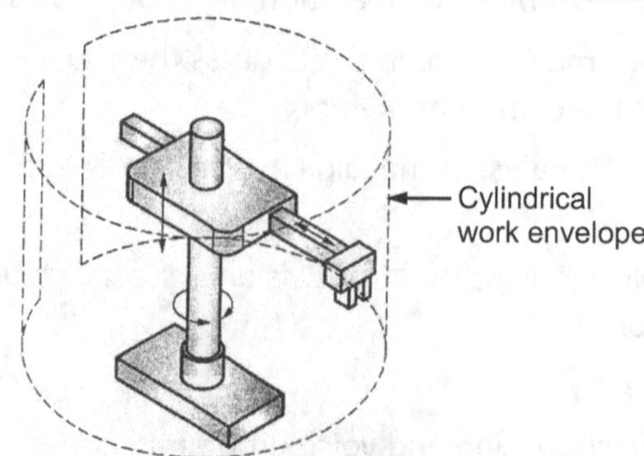

Fig. 6.19: Cylindrical Configuration Robot

- Cylindrical configuration robot shown in Fig. 6.19, provides two linear and one rotary motions.

- This configuration, which provides cylindrical work envelope, has good work area to floor area ration.

- Cylindrical configuration robots are used for loading and unloading of machine tools.

6.8.3 Polar (Spherical) Configuration Robots

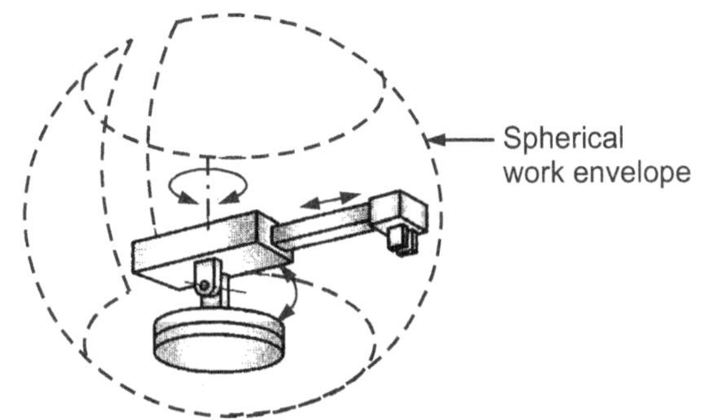

Fig. 6.20: Polar (Spherical) Configuration Robot

- Polar (spherical) configuration robot, shown in Fig. 6.20 provides one linear and two rotary motions.

- This configuration provides spherical work envelope.

- Polar (spherical) provides configuration robots are used for spot welding and manipulation of heavy loads.

6.9 APPLICATION OF ROBOTS

This applications of robots in manufacturing industry are broadly classified into five areas.

6.9.1 Machine Loading and Unloading

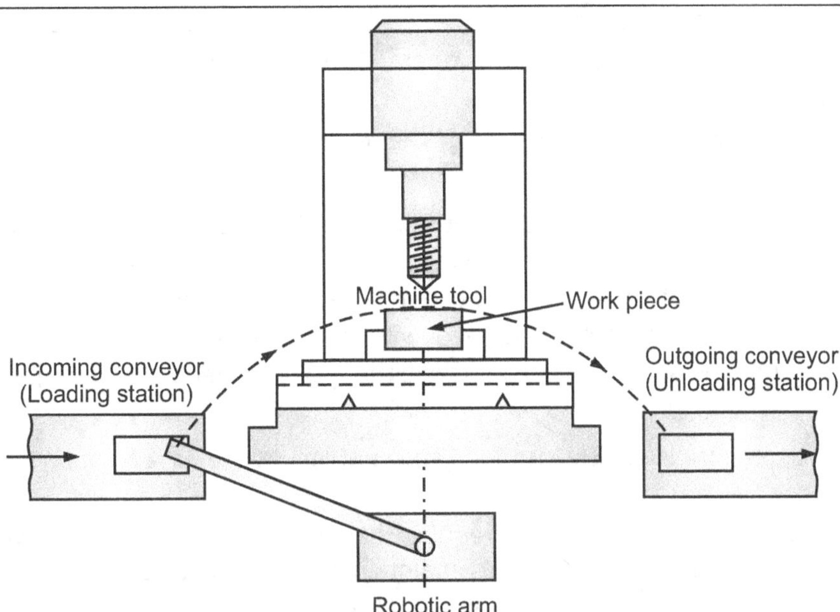

Fig. 6.21: Robot machine Loading and Unloading

- Robots are used for loading and unloading of parts in CNC machining centers. Flexible manufacturing systems die casting machines, punching press etc.

- The use of robots in such machines reduces the part handling time, thereby reducing the cycle time and hence improving the productivity.

- In machine loading and unloading, a robot should be able to orient the workpiece correctly so as to locate it accurately as a machine after picking it from bins or conveyor.

- Fig. 6.21 shows the use of robot for loading and unloading purpose in flexible manufacturing system (FMS).

6.9.2 Material Handling

- The robots are used for shifting the material or finished parts machines, conveyor, or feedor to the storage pallets and arranging them in order, as shown in Fig. 6.22 such operation is known as palletizing.

- The robots are also used for shifting the material from storage pallets to the machine conveyor or feedor. Such operation is knows as depalletizing.

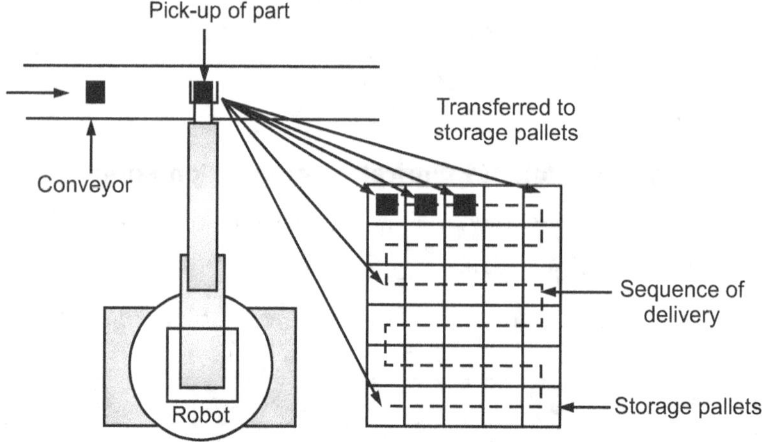

Fig. 6.22: Robot Material Handling

6.9.3 Processing Operations

- The robots are also used for performing the operation on workpiece. The robots are equipped with tools mounted on the end effectors.

- Some of the processing operations performed by the robots are discussed below.

(i) Spot welding:

The spot welding operation is widely used in automobile industries for car body manufacturing. The use of robot for spot welding not only reduces the manpower requirement drastically but also improves the quality as well as the rate of production. The point-to-point (P-T-P) servo controlled robots (either hydraulically or electrically actuated), equipped with spot welding gun, are normally used for this purpose. The configuration used for such robots is either polar (spherical) type or jointed-arm type.

(ii) Arc welding:

The arc welding operation is widely used in automobile industries and manufacturing of process equipment.

The arc welding operation often involves irregular shaped seam. In manual arc welding operation of irregular shaped seam, it is really difficult to maintain the continuity of run as well as uniformly of strength and throat thickness. The arc welding operation requires high labour skill. In addition, the working conditions are hazardous.

Hence, robots are preferred in the mass production type applications of arc welding. The use of robots improves the quality of welding as well as rate of production. The continuous path (cp) servo-controlled robots with polar (spherical) or jointed-arm type configuration are used in mass production applications of arc welding.

(iii) Spray painting:

In spray painting, a fine mist of paint (both lead and plastic based) is carcinogenic. It is highly hazardous to human health. Therefore, the modern paint shops use robots for spray painting operation. The continuous path (cp) servo-controlled robots with polar (spherical) or jointed-arm type configuration are used for spray painting with the use of robots, the resultant coating is for more uniform than a human being can produce. This results in a higher quality product and less consumption of paint.

(iv) Machining operations:

The robots are used for different machining operations like: milling, drilling, grinding etc. the rotating spindles are used as end effectors. The tools are fixed to the rotating spindles for performing the machining operations. The robots used for machining operations are subjected to high cutting forces, and hence should be robust. The Cartesian configuration and cylindrical configuration robots are widely used in machining operations.

6.9.4 Assembly

- In assembly two or more components are added to form a new entity. The assembly normally involves following operations:

  1. Mechanical fastening,

  2. Soldering and brazing,

  3. Welding,

  4. Press fitting, and

  5. Adhesive bonding.

- The assembly involves highly repetitive and boring operations which lead to human fatigue. This may adversely affect the product quality and productivity. The use of robots in assembly results in reduction in manufacturing cost and improves the productivity.

- The most commonly used configuration for assembly in SCARA robot.

6.9.5 Inspection

- Robots are used for inspecting parts or sub-assemblies.

- The inspection probes mounted on the end effectors are used for checking the dimensions. The checked dimensions are compared with the predetermined values.

- In some cases, the robots separate the rejected parts.

Important Points

1. Robot:

- A reprogrammable, multi-functional manipulator designed to move materials, parts, tools or specialized devices through variable programmed motions for performing a variety of tasks.

2. Robotics:

- The combination of machine tool technology and computer science and is a form of industrial automation and is a technology with a future and for a future.

3. Basic components of robots:

(a) Manipulator	(b) Controller
(c) End effector	(d) Processor
(e) Sensors	(f) Actuator
(g) Software	

4. Robot joints:

(a) Linear joints	(b) Orthogonal joints

    (c) Rotational joints   (d) Twisting joints

    (e) Revolving joints

5. Actuators:

- Actuators are the devices which provides the actual motive force for manipulators joints of the robots.

- Type of actuators:

    (a) Mechanical actuators   (b) Hydraulic actuators

    (c) Pneumatic actuators   (d) Electric actuators

6. End effectors of robot:

- End effector is a device that is attached to the wrist of the robot arm so as to enable the robot to perform a specific task.

- Type of end effectors:

    (a) Grippers   (b) Tools

- Gripper are the end effectors used for holding the parts or objects.

- **Tools:** In many applications, robot is required to operate tools rather than handling the parts. In such case, tools are used as the end effectors.

7. Robot sensor:

- Robot sensors are used to estimate a robots condition and environment. These signals are passed to a controller to enable appropriate behaviour.

- Sensors in robots are based on the functions of human sensory organs.

- Robots require extensive information about their environment in order to function effectively.

8. Basic configuration of robot:

- Base on the co-ordinate system of motion of the manipulator and end effector, there are three basic configuration of robots.

    1. Cartesian configuration robots.

    2. Cartesian cylindrical configuration robots.

    3. Configuration robots.

    4. Polar (spherical).

9. Application of robots:

    (a) Machine loading and unloading.

    (b) Material handling.

    (c) Processing operation.

        (i) Spot welding.

        (ii) Machining operation.

        (iii) Spray painting.

        (iv) Arc welding.

    (d) Assembly.

Practice Questions

1.   Definition of robot and robotics.

2.   Advantage and disadvantage of robotics.

3.   Explain basic component of robot.

4   Type and Explain robot joints.

5.   Explain degree of freedom of robot.

6.   Explain different type of actuators.

7.   Explain mechanical actuators.

8.   Explain hydraulic actuators.

9.   Explain pneumatic actuator.

10. Explain electric actuator.

11. Explain different type of end effectors.

12. Short note on robot sensor and classification of sensor.

13. Explain basic configuration of robot.

14. Explain application of robot in loading unloading.

15. Short note on application of robot in material handling.

16. Short note on application of robot in processing operations.

17. Short note on application of robot in assembly and inspection.

MODEL QUESTION PAPER - I
(On Unit 1, Unit 2, Unit 3)

Marks: 20   **Time:** 1 Hour

Instructions:

1. All questions are compulsory.
2. Illustrate your answers with neat sketches wherever necessary.
3. Figures to the right indicate full marks.
4. Assume suitable data if necessary.
5. Preferably, write the answers in sequential order.

Q.1. Attempt any Four: **(08)**

(a) Draw a block diagram of traditional product life cycle.

(b) State and explain the benefits of CIM.

(c) What is concurrent engineering?

(d) State the types of CAPP.

(e) Classify the CIM network.

(f) Write note on: Bus network.

Q.2. Attempt any Three: **(12)**

(a) Explain with block diagram, CAD/CAM/CIM product cycle.

(b) Explain with neat sketch, systems of CIM.

(c) Explain analysis, optimization and evaluation for the given part using CAE.

(d) State and explain the types of database.

(e) Draw a typical hardware layout of CIM system.

✍ ✍ ✍

MODEL QUESTION PAPER - II
(On Unit 4, Unit 5, Unit 6)

Marks: 20   **Time:** 1 Hour

Instructions:

1. All questions are compulsory.
2. Illustrate your answers with neat sketches wherever necessary.
3. Figures to the right indicate full marks.
4. Assume suitable data if necessary.
5. Preferably, write the answers in sequential order.

Q.1. Attempt any Four: **(08)**

   (a)   Define: Group technology and cellular manufacturing.

   (b)   State any four limitations of automation system.

   (c)   Draw the layout of FMS system.

   (d)   Define: Industrial robot.

   (e)   Draw a block diagram of : Rotary layout type FMS.

Q.2. Attempt any Three: **(12)**

   (a)   Explain with neat sketch, any two types of machine cells.

   (b)   Explain with neat sketch, different joints of robot.

   (c)   Compare different types of automation systems.

   (d)   Explain the main elements of automation system.

   (e)   Write note on: Electric actuators used in robots.

☝ ☝ ☝

MODEL QUESTION PAPER - III
(End Semester Examination)

Marks: 70   **Time:** 3 Hours

Instructions:
1. All questions are compulsory.
2. Illustrate your answers with neat sketches wherever necessary.
3. Figures to the right indicate full marks.
4. Assume suitable data if necessary.
5. Preferably, write the answers in sequential order.

Q.1. Attempt any Five: **(10)**
   (a)   Define automation. State three examples of automation.
   (b)   State any five advantages of flexible manufacturing systems.
   (c)   Write note on: Database Management System.
   (d)   Define supply chain management. State the steps in supply chain.
   (e)   State any five benefits of CIM.
   (f)   State the stages of PLM.
   (g)   State the elements of automation system.

Q.2. Attempt any Three: **(12)**
   (a)   State and explain the advantages of database management system.
   (b)   Classify flexible manufacturing system.
   (c)   What are the different strategies in automation?
   (d)   Explain with block diagram, Programmable Logic Control.
   (e)   Explain the use of magnetic grippers in robot. State their advantages and limitations.

Q.3. Attempt any Three: **(12)**
   (a)   Explain with block diagram, LAN.
   (b)   Compare the two types of manufacturing systems.
   (c)   Explain with neat sketch, elements of CIM system.
   (d)   Explain any one part classification and coding system.

Q.4. Attempt any Three: **(12)**
   (a)   Differentiate between: Variant CAPP system and Generative CAPP system.
   (b)   Explain with neat sketch, any two methods of grouping parts into part families.
   (c)   Explain with neat sketch, basic components of robot.
   (d)   Justify the need of automation.

Q.5. Attempt any Two: **(12)**
   (a)   Explain the types of networks and network topologies with diagram.
   (b)   Explain with neat sketch, variant CAPP system.
   (c)   Explain with neat sketches, Cartesian and cylindrical configuration robots.

Q.6. Attempt any Three: **(12)**
   (a)   Explain the types of automation.
   (b)   Explain the procedure for computerized part program generation for a given part.
   (c)   Explain any two applications of robot.
   (d)   State and explain the needs of modern production system.

☞☞☞